R!SKS &
DETERRENTS
IN CONSTRUCTION
PROJECTS

For Customers, Contractors, Suppliers & Consultants...

PAWAN DUA

INDIA • SINGAPORE • MALAYSIA

Notion Press

No.8, 3rd Cross Street,
CIT Colony, Mylapore,
Chennai, Tamil Nadu – 600004

First Published by Notion Press 2021
Copyright © Pawan Dua 2021
All Rights Reserved.

ISBN 978-1-63669-722-2

This book has been published with all efforts taken to make the material error-free after the consent of the author. However, the author and the publisher do not assume and hereby disclaim any liability to any party for any loss, damage, or disruption caused by errors or omissions, whether such errors or omissions result from negligence, accident, or any other cause.

While every effort has been made to avoid any mistake or omission, this publication is being sold on the condition and understanding that neither the author nor the publishers or printers would be liable in any manner to any person by reason of any mistake or omission in this publication or for any action taken or omitted to be taken or advice rendered or accepted on the basis of this work. For any defect in printing or binding the publishers will be liable only to replace the defective copy by another copy of this work then available.

Contents

Acknowledgements   *5*

Introduction   *9*

Chapter 1 Understanding Projects & Risks   15

Chapter 2 Project Selection Risks   29

Chapter 3 Partner Selection Risks   39

Chapter 4 Financial & Commercial Risks   55

Chapter 5 Scope & Technical Risks   77

Chapter 6 Supply Chain Risks   98

Chapter 7 Quality Risks   123

Chapter 8 Execution Risks   135

Chapter 9 Health, Safety & Environment Risks   160

Chapter 10 Legal & Compliances Risks   182

Abbreviations   *207*

Acknowledgements

I am blessed to have an outstanding group of people in my network, who have constantly guided and supported me to accomplish this book so that I can contribute to strengthen the risk management practices in the construction sector. It is important to defend the interest of the parties involved in managing their businesses, relationship, and goodwill.

Without a significant contribution from many experts, this book would have not been a reality. It is attempted to cover the lessons related to recurrent risks and deterrents experienced in the construction sector that affects the projects and levers adopted by project experts to manage them.

It is difficult to capture all the motivators who supported, guided and mentored to write this book. Indeed, the list is long, yet it would be injustice if I miss expressing my special gratitude to the major contributors to this book.

- My Family—**Vipla** (my Wife), **Anchal** (my Daughter) and **Sahil** (my Son) who have been my true source of energy, motivation and inspiration to create this legacy.
- **R. Narasimhan Iyer,** Director (Operations), SB Energy (Softbank Group), has been my *guru* and mentor over the last 18 years, who guided me across various projects and service

deliveries. He has been instrumental in guiding me through various aspects, complexities, risks and deterrents in projects. I am deeply indebted to his significant contribution in making this book a reality. His approach of constant development and evolution for self and his team is outstanding, of which I am one of the privileged beneficiaries.

- *Jayanta Guha,* Regional Head (MEP Business – South & East India), Sterling & Wilson, my friend, mentor, and a seasoned professional. He shows commitment and perseverance in exploring different ways of doing complex projects and evaluating risks with pragmatic possibility of its occurrence and consequence. His approach of evaluating risk without any emotions, biases, or verbal assurance is a great driver of thoughts.

- *Dipen Khandeliya,* Head (Engineering & Asset Management), Sprng Energy, my friend and colleague, who is a true technical expert and openly shared technical knowledge, wisdom and nuances of construction projects. His approach is simple and straight, which helped me to put the technical and scope perspectives in the easiest possible manner in this book.

- *Yogesh Srivastava* a veteran in Health, Safety, Environment and Sustainability. He truly motivated and supported me to enrich the contents of HSE, while making this as an integral part of project management and an overall organisational fabric.

- *Vivek Choudhary,* Senior Legal Counsel, Schneider Electric, a sound and deep analyst of legal, statutory and compliance aspects of projects. His approach of detailing and envisioning counterparty reaction is of great support in identifying and managing risks in projects.

- **Vivek Gupta,** Associate General Manager (Techno-commercial), Adani Transmission, an experienced expert in managing the supply chain for large projects, who has contributed in a variety of ways to enrich the contents of my book and shared domain experiences for wider benefit.

- **Arunava Sengupta,** Solution Risk Manager, Schneider Electric, greatly supported me in identifying a variety of risks, which can trigger at any stages of the project and may go unnoticed with project team. As such, these risks can significantly impact commercial, contractual, or legal aspects projects.

Many people have shared their pearls of professional wisdom, experiences and learnings, which proves that risks are universal, integral, and inevitable part of the projects. The only way to deal with risks is to inculcate the culture of risk management and implement the mechanism to manage them effectively and efficiently.

Introduction

Construction projects are subject to higher risks, uncertainties and deterrents as compared to other industries or businesses. The prime reason for high risks could be the diverse scopes, activities, processes, resources, environments, and organisations' complexities involved in projects. As most of the large and complex projects are executed on a turnkey basis, the contractors take the major chunk of risks and responsibilities. The risks start from the selection of the right project, right customer, right partner, conceptualisation of the scopes or assuming responsibilities.

Changes are inevitable in projects and so are the risks. These are integral parts of the projects and cannot be viewed independently and dealt with without a complete understanding of the project's scopes and deliverables.

Projects can have risks related to scopes, technical, technology, schedules, financials, quality, supply chain, legal and compliances etc. In projects, the impact of risks could be minor to devastating. There is no standard rule or guideline which can mitigate all the risks in projects, despite replicating the same design, resources, tools, methods and technologies. It is also a myth that by deploying the best experts and resources, an organisation can zero down the risks in the project. However, with professional and matured risk management processes, these can be mitigated in the best possible way and curb losses, damages, and disputes.

Projects are unique, as they differ in one or more parameters related to the scope, sites, schedule, resources, dependencies, conditions, or environment under which these are executed. Such uniqueness makes the projects risky and adds risks to the projects. Project risks can adversely affect the goals or objectives. Consequently, such outcomes can not only demotivate stakeholders but also attract financial or goodwill losses.

Project risks occur when deliverables are affected with unexpected events or conditions, thereby derailing the forecasted timeline, budget, scope, quality, or the objective itself.

Projects can fail, cause losses or doom due to dozens of reasons. Though project teams can recite a long list of reasons, justifications or causes of such risks which have the ability to doom, fail or cause losses to the projects. Even the best-planned projects can have unexpected risks related to:

- **Land** – selection, acquisition, right-of-way, encumbrances, disputes, or litigations.
- **Grids & transmission networks** – readiness, allocations, restrictions, or unavailability.
- **Approvals & permissions** – long and complex processes, rules, guidelines, documentation involving multiple stakeholders private and government agencies.
- **Material costs** – currency and commodity fluctuations, besides availability, delivery and cost.
- **Logistics** – availability, mode of vehicles, routes, clearances, weather, strike, documentation, customs and exemptions.
- **Funds & Payments** – Delays or disputes in receipt of timely payments.

- **Resources** – unavailability of adequate manpower or construction equipment.
- **Force majeure** – Including extreme weathers can impact performances and productivity.
- **Change in law** – Government rules and laws can change anytime.
- And many other known or unknown risks.

Thus, risk management process is required to identify and manage the potential risks and problems that can endanger the project. All such risks require incessant assessment, analysis and appropriate actions or mitigations to avoid damage once occurred.

Overall risks in any project are always higher than the risks covered in the risk register or as perceived by its stakeholders. Unless organisations are matured and rigorously follow professional risk management processes, risks can continue to hit the projects and businesses severely. Risk management in construction projects is essential to minimise the losses and enhance the chances of improving profitability. Thus, the risk register does not reflect the true and complete picture of project risks. Additionally emergent risks and uncertainties, which are neither foreseeable nor predictable can also affect the project materially.

For all practical reasons, one cannot identify all the risks and document them in the risk register. As the project progresses, risks keep emerging and it makes the risk management an iterative process which is followed over the project lifecycle. In order to handle risks effectively, the project team should know the level of risk exposure and risk appetite of the organisation, so as to manage the same in the most effective and efficient manner.

The purpose of this book is to guide and support professionals, aspirants and students who are practising or willing to make their career in construction management, risk management or project management. Throughout the book, emphasis is given on the risks and deterrents commonly experienced by the construction fraternity and strategies applied to overcome the same and protect their project or business interests.

Predominantly, this book covers the interest and perspective of various types of contractors, suppliers, consultants and service providers. This may include MEP, EPC contractors, sub-contractors, equipment and material suppliers, consultants, consortium partners, and other associated services providers.

Contractors or suppliers may use different types of contracting structures or models like composite, divisible, lump-sum turnkey (LSTK), BoQ-based, framework agreements, etc. However, managing risks are critical and crucial for the success of the projects as well as the organisation.

The terms 'contractor,' 'supplier' and 'service providers' are used interchangeably for the sake of convenience. However, all represent the same party, meaning and carrying huge vistas of risks, responsibilities, and liabilities.

At times, a contractor may participate in tenders or opportunities along with other partners in a consortium arrangement, which bring radically a different set of risks, challenges, issues and liabilities. The risks associated with consortium partners and mechanism have also been touched upon in this book so that the interests of all consortium parties are protected.

A genuine attempt is made to translate actual experiences and challenges into this book, rather than theories, which may be less practised. The author has tried to migrate thoughts and experiences from *"Mastak"* to *"Pustak"* (Brain to Book), so that project practitioners can easily resonate with their own project environments, situations, or experiences.

It is important to understand that the risk management is not the sole responsibility of sales managers, tender managers, project managers, legal counsels, contract managers or risks managers, but is a collective responsibility of wider teams, groups, or organisations. Involvement of complete project team is important to develop the sense of ownership, responsibility and accountability in identification of the risks and working out the suitable risk response actions collectively.

Each team member has a specific role and purpose in the project and organisation. No project can succeed until all the relevant stakeholders and team members have fair inclination to support and perform their respective duties and obligations diligently commensurate to their roles, function or domain in the projects or organisation.

Many project personnel are unable to differentiate risks with requirements, issues, or problems. Rather, they interchangeably use or populate their risks register with issues, requirements, or problems. Whereas the actual risks are about *future uncertainties.* Thus, the risk register should reflect future events which matter most to the project.

This book is expected to serve as a bible or reference guide for many new and practising project professional or aspirants. This shall help them to develop the culture of risk management in their organisation and carry out 360-degrees risk management of each project before deciding whether to pursue, abandon or drop the project.

Lastly, risks may occur at any stage – starting from inception, selection, bidding, planning, designing, execution, construction, development, closure, or operation phases of the project. So watchful risk management helps to decide what can be absorbed and what can be avoided. Promoting a risk-awareness culture can help to set up a strong defence layer against any risk and respond if any risk surfaces and threatens the project or organisation.

Chapter 1

UNDERSTANDING PROJECTS & RISKS

"Risk comes from not knowing what you're doing"

– Warren Buffett

What is a Project? As per PMBOK Guide version 6.0, *"A Project is a temporary endeavour undertaken to create a unique product, service or result. Projects are undertaken to fulfil objectives by producing deliverables... Deliverables may be tangible or intangible... Projects are undertaken at all organisational levels. A project can involve a single individual or a group. A project can involve a single organisation unit or multiple organisational units from multiple organisations."*

However, the exact definition of a project may vary for different organisations, depending upon their business model, requirements, manageability, or structure, etc. Some can define the project as simple as a supply of material or equipment (single item or a few quantities) to as complex as full-fledged construction of infrastructure like buildings, industrial plants, highways, or such large facilities or contraptions.

Different organisations execute projects, which may include comprehensive or limited scopes like:

- Design, engineering, manufacturing, supply, (civil, mechanical & electrical works) construction, erection, testing, commissioning and hand-over of complete facility

- Supplies of workforces or consultancy services

- Operation and maintenance of plants, facilities, or infrastructure

- Develop or revamp systems, structures, processes, or organisation

- Design, develop, integrate, customise, or enhance system software or application

- Develop and diversify in a new niche, segment, markets, regions, countries, etc.

Generally, construction projects are executed with higher risks and responsibilities. Most large contracts are preferred and performed on a turnkey or composite basis thereby transferring higher risks, rewards and liabilities to the contractors. Such projects involve diverse scopes, risk and responsibilities for the complete design, engineering, procurement, manufacturing, supplies, logistics, construction, erection, testing, commissioning, operations, etc. These projects may involve the supply of various material and equipment, deployment of large workforces, specialised consultancies, softwares, integrations with existing systems or plants or networks.

Large projects may have a crucial dependency upon gig economy, political systems, sizeable land banks, readiness of grids, transmission systems, vast supply chain, fund availability and many other compliances and permissions. These activities not only involve high risks but may also heavily impact project rigorous plans and objectives. Thus, the responsibility and accountability of such activities must be explicitly defined and agreed between parties so as to complete the projects on time and meet the objectives. If there is lack of clarity or shared accountability, there is a high probability of conflicts or avoidance of responsibility.

Construction projects mainly carried out either on 'green-field' or 'brown-field' basis.

- *Green-field* projects are the construction of new buildings, plants, or other facilities and are carried out on barren, waste or agricultural land, post-conversion of land utility. Work on such areas is done for the first time with initial land development. Such projects generally do not require demolition or remodelling or renovation of the existing infrastructure or plants. However, existing hurdles may need to be removed or diverted like telecom, water, gas or electric lines, pathways, tube-wells, tombs, crops, etc.

- *Brown-field* projects are related to upgradation, renovation, revamping, expansion or augmentation of any existing infrastructure, facilities, or plants, etc. Such projects may also include demolition, remodelling or redevelopment of the existing setup, infrastructure, plants, machinery and other systems.

Project life cycle (PLC) definition may vary from organisation to organisation. Some organisations can define the start of a project from the identification of a new lead or opportunity, or launch of tender, or upon winning an award, or split a large project into multiple sub-projects suiting to the business units, functions, structures and processes. Many organisations assume project life cycle by getting along with potential customers in defining scopes or specifications thereby influencing or enabling them to make prescription or tender formation, which are more conductive for them. By doing so, these organizations may incorporate the specifications or conditions, which are more favourable to them and increase their probability of winning.

Projects can be of multiple combinations like (a) minimum resources of a function, (b) multiple resources from multiple functions, (c) involving multiple organisations, locations, cross-countries, etc.

Project Lifecycle in an organisation may look like this:

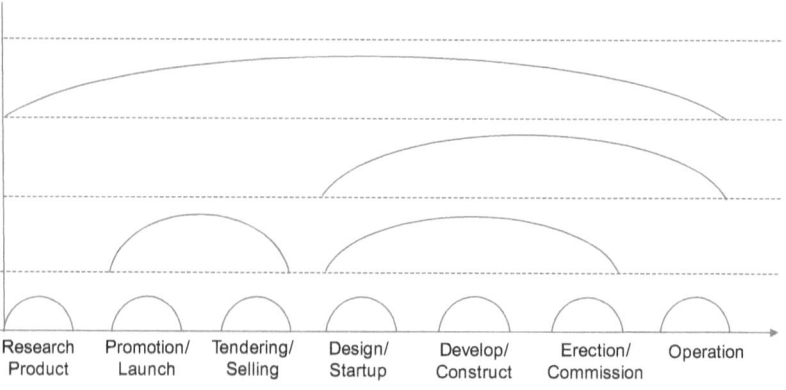

Projects can be as simple as supplying a tailor-made product or as complex as setting up a new industry or plant or even bigger customized contraption.

In case of multi-company projects, cross-functional experts from all organisations can collectively be involved and contribute to identify project deterrents and risks and find suitable mitigation plans to achieve win-win situations.

What is Risk? There are plenty of questions with numerous definitions, understandings, theories, and opinions about risks. In a true sense, a risk is an ***uncertain futurist*** event or condition, or set of events or conditions which can impact, negatively or positively, the project baselines or objectives.

It is commonly perceived that the project baseline is limited to project timeline schedule. Whereas project baseline is not limited to project schedule only, rather all the basis, conditions and assumptions agreed at contract-formation become the baseline. This include scopes, specifications, deliverables, schedules, performance parameters, obligations of parties, warranties and representations, assumptions, and other agreed provisions. This also covers the prevailing laws and regulations, standards, taxes and relaxations as applicable and agreed at the time of entering into the contracts. Any changes in baseline can impact one or other conditions of the project, regardless of who owns the risk and reward, as a consequence of such change.

Ironically, some professionals assume that risks are only negative and should be avoided or neglected. Irrationally avoiding, hiding, or escaping risks may breed new secondary risks, sometimes. For example, it is safe to stay at home to avoid the risks of accidents, but by doing so, we may lose many opportunities for earning, enjoying, or exploring the world. Rather than neglecting or avoiding, the best way is to manage the risks.

Project risks signify the likelihood or potential of a future event. The world is full of uncertainties. Risk is a sub-set of uncertainty. However, it does not mean that all uncertainties of the world are risks to your project. Hence, knowing fully well the content and context of the project is far more important than knowing the uncertainties or risks in isolation. Risks are those futuristic uncertainties or occurrences, which matter the most to the project.

Fundamentally, risks have two components – (1) *Uncertainty* – may or may not occur, and (2) *Impact* – positive or negative. So, any event which if occur may impact project scope, time, cost, or scope (objective/deliverable) is a risk.

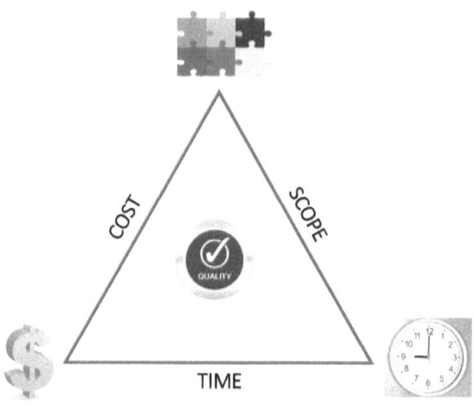

Uncertainties may impact the projects differently, depending upon the stage, complexity and mitigation plan. Some of the scenarios include:

Scenario – 1: **One common risk – Affecting multiple projects**

Unknown risk of *force majeure* event like recent outbreak of pandemic Covid-19 coronavirus had impacted most of the ongoing construction projects for most organisations in infrastructure sector.

Scenario – 2: **Risk of currency fluctuations – No Impact**

In a project that involves incomes and/or costs in local currency within the country, the fluctuations of foreign exchange rates may not impact project directly, as all the materials and services are manufactured or available within the same country.

Similarly, in a project that involves incomes and/or costs in currencies (other than local currency), if the currency fluctuation occurs at the fag end of the project when all purchases are completed, it may also not impact the project even if the project involves the multi-currency transactions.

Scenario – 3: **Risk of currency fluctuations – Impact**

If a project involves the import or export of machines, equipment or resources from other countries, it may involve transactions in multi-currencies (other than local currency), and the foreign exchange fluctuations can directly impact. For example, the de-valuing Indian Rupee versus the US dollar can have a significant impact on the project financials. Now, this impact can be negative, whereas, if the project involves export from India, the project could have a positive impact from the same risk.

So, considering the timing and event of the risks, the impact could be ascertained and managed accordingly.

Project people may encounter a variety of risks, which can be categorized as under: –

- **Known-Known risks** – The *project team already knows about it*. These are not the risks for projects and are suitably factored as part of scope, costs, facts, or requirements.

- **Unknown-Known risks** – The *project team does not know but others know about it*. These are hidden facts but someone in the team, stakeholder, organisation or community knows such risks. These are untapped knowledge, learning or wisdom. These cause regular and frequent changes or variations that may swing the planned productivity, schedule, budget or objective of the project.

- **Known-Unknown risks** – The *project team knows but others do not know or ignore*. These risks are classic risks, which are identifiable, predictable, and foreseeable and yet have rare exceptions. These types of risks may be envisioned or anticipated by project teams as the probability of occurrences (intentionally or unintentionally) or the likelihood of their effect is known.

- **Unknown-Unknown risks** – Neither *project team nor others know about it at all*. These are unforeseeable or unpredictable risks and whose probabilities of occurrences and effects cannot be assumed or considered even by the most experienced experts. These are usually considered as *force majeure or advance technological inventions*. Traditionally, such risks were considered out of risk management scope.

What is Risk Management? Risk management is not about hiding or stopping the risks or predicting the future. It is about understanding the risks and taking informed decision by appropriately mitigating it in the most effective way. Stakeholders may use different levers to accept, transfer, escalate, avoid, or enhance the risks.

If the risks are beyond acceptable limits or risk appetite of the organisation, the stakeholders can even decide to terminate, abandon or shelve the project to save time, cost, resources and avoid further losses or bigger surprises.

As per PMBOK Guide version 6.0, *"Project Risk Management includes the processes of conducting risk management planning, identification, analysis, response planning, response implementation, and monitoring risk on a project. The objectives of project risk management are to increase profitability and/or impact of positive risks and to decrease the probability and/or impact of negative risks, in order to optimise the chances of project success."*

Despite adhering to the best practices or processes, one cannot guarantee that all risks are identified and mitigated any point of time during the project life cycle. As risk management is an iterative process, stakeholders must be watchful throughout the project lifecycle. It is also a myth that by implementing risk management or by deputing experts in projects, one can completely eliminate or bring down all the risks to zero-level.

People make a difference. This connotation can have positive or negative impacts. Many people tend to hide, ignore or underestimate the probability or impact of risk and hence abstain from including them in the risks register, whereas, by doing so, they lose the opportunity of a due analysis and may confront the consequence later. It is always advantageous to detect the risk as early as possible, include them in the risk register and plan its management before it turns bigger, bitter, or lethal. It is also not uncommon that many organizations completely ignore risk registers or assume that risks are limited to risk register.

In construction projects, there is a high likelihood of risks occurrences due to events, factors or combinations, which can be detrimental to the projects due to lack of predictability or weak risk-response strategy. Outcomes or results of risks could be better or worse than expected. Broadly, project risks can be categorised as – internal and external. Various types of risks are broadly reflected in the following graphical form.

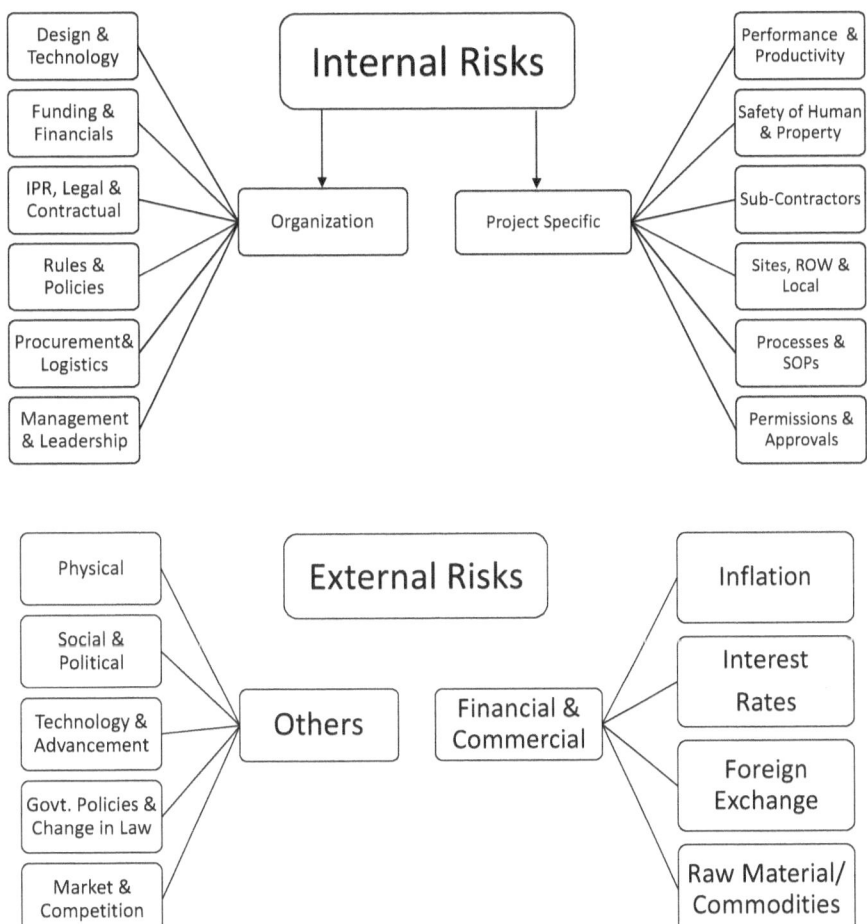

Although, the probability of any risk can be anything between 1% to 100%. However, while analysing the risk, some users may not carry out a detailed assessment. For sake of convenience or due to weak understanding of the risk, some professionals assign the probability of risk as 0%, 50% or 100%. Such guess reflects the poor understanding or analysis of the risk. Any risk which assumes a 50% probability means, it needs more analysis or clarity to define the risks correctly and similarly any risk with zero percent probability may not be a risk to the project

Depending upon its risk impact – 'negative' or 'positive' – the strategy of response to the risk may vary.

The focus of risk management is broadening so as to ensure that all type of risks are considered and covered during the risk management cycle, which may include the following:

a. **Event Risks** – Occurrences that can impact the smooth functioning of some activities or complete project. Event risks are like – pouring of heavy rains at the construction site, labour strikes, customer requirements can change, key supplier can go out of business, critical resource can quit the job, unplanned power shut-down, etc. Such event risks can impact time, cost, or quality of the project.

b. **Variability Risks** – It is contrary to event risk or may be considered as non-event risk. This results in the potential spread of possible values from some parameters related to planned event or activity. Outcomes can be less or more than planned. It is like a game of chance with dice, where there are several possible outcomes. For example, the number of defects identified in a field quality inspection may differ depending upon Inspectors, productivity can go up or down due to inclement weather conditions or other factors.

c. **Ambiguity Risks** – While conceptualising or bidding for a project, many inputs or data are assumed, due to the absence of accurate or complete quantitative information, knowledge, and understanding. These assumptions may undergo changes with the development of work or passage of time; for example, assumed foreign currency rates, interest borrowing rates, market, or competition dynamics, etc. Such risks are managed through exploration, extrapolations, or experiments. Similarly, organisations may plan prototypes, simulations, or sample approvals before taking up mass production or performance testing. This is an authentic situation in which the ambiguity risk is present and can be encountered in the full or mass lots.

d. **Emergent Risks** – (Unknown-unknown risks) – These risks emerge from outside of human experience, mindset, or wisdom. These are caused due to limitations of the human conceptual framework of worldview. These risks are entirely outside the radar of the project stakeholders. Outcomes can be unknown, unknowable or disruptions. Thus, it is impossible to give an exact example, though such risks can be game-changers or paradigm shifters. These can be disruptive or catastrophic for projects or businesses. These risks may result in future due to technological advancements, regulatory frameworks, inherent project complexities or VUCA environment.

Initially, organisations may not have sufficient quantitative information related to risks. So the risk exposures normally look high or low, but as the time goes, team's performance analysis, information and situations become relatively clearer, organisations evaluate their feasibility and then take suitable corrective or mitigation actions.

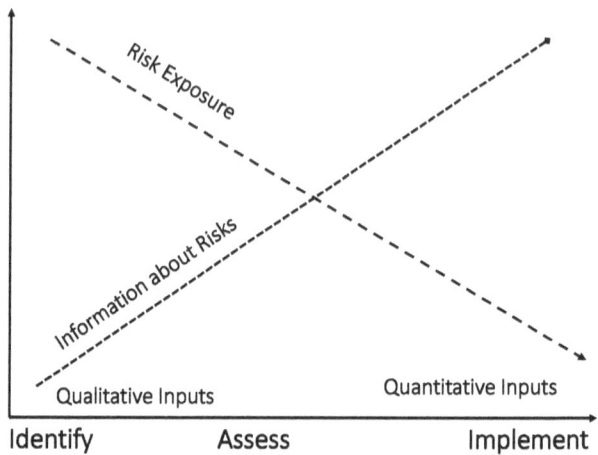

Risks are associated with future events, which have not happened yet, whereas if an event has already occurred, it is considered an 'issue.' Many people mix *risks* with issues, requirements, constraints, or problems. The project team has to clearly ensure that the risk register carries only the risks – the uncertainties. Whereas separate issue logs or delay logs should be maintained by the project team and timely actions should be taken to complete the issues or overcome the delays.

For example, if any contract requires that an equipment should have valid type tests conducted in last five years, at given voltage, rating, condition, standard or limits but you do not possess the same, it does not mean that it is a risk. It is a clear, known, defined requirement and the same must be complied with. Such a requirement should be duly assessed in the beginning and adequate provision (cost and time) be envisaged. However, it may be considered as a risk if that specific equipment has failed previously during type tests or such equipment has never been type-tested but incorrectly committed to the customer. In such situations, the organisation should plan whether the alternate equipment can be pitched, or type tests are completed before binding itself in contractual relationships.

Likewise, if a project requires specified numbers of certified professional to be deployed at the site to meet the scope requirement, it is still the project requirement. In case an organisation does not have the certified professionals, it must hire within the available time-frame. Notwithstanding, there are always risks associated with people, related to their full-time availability, competency, attitudes, other individual preferences or concerns.

Not to forget that many risks may occur multiple times during project lifecycle. So, the project team needs to be continuously vigilant about risk management process. For example, in a long duration project, change in law or tax structures can occur multiple times. Social commotions or labour strikes can cause work to stop multiple times. Poor workmanship or untrained manpower can create repeated defects causing reworks or rejection of works.

Some contracts require obtaining government approvals, permissions, or certifications, which involve fees, charges, or liaison with authorities. If the contractor missed considering the appropriate time and cost towards statutory approvals, this can impact the project legally and financially, or even cause project delays. For example, some tenders or applicable laws require registration of the contract with particular jurisdiction or authorities, which may involve fees payments. If such fees are not considered, it will hit the project financially.

Customers may ask contractors to use specific tools, software, methodologies, or processes like project documents to be shared in MS Projects or Primavera, AutoCAD, E-Tab inventory, ERP-based invoices, or other root-cause-analysis or decision-making methodologies, etc. Such requirements must be carefully envisaged and planned by the contractor, to avoid unnecessary conflicts or risks at a later stage.

With rapidly growing complexity in the construction segment, the need for risk managers, contract managers and legal counsels have increased in the last decade or so. These experts not only bring tangible but intangible values to the businesses, whose value cannot be assessed in a short time. Engaging these experts also could help to reduce revenue and profit leakages, enhance stakeholders' confidence and trust, and increase the probability of success for any project or organisation. If the contracting organisation does not have in-house experts, it can explore to hire or leverage services from external agencies.

To conclude, risks are inevitable in projects. Risks bring thrills and threats to any project. Risks can be negative or positive. Regardless of the measures adopted, it cannot be zero-down. Project teams should adhere to the risk management process and rigorously follow it from inception until close-out of the project. Some risks can turn into opportunities, which an organisation must explore, enhance and exploit. Some of the risk management strategies adopted by project teams include – *(a) accept or remove risk, (b) reduce likelihood and impact, (c) transfer risk, (d) make a contingency plan, or (e) explore or enhance positive risks.*

Chapter 2

PROJECT SELECTION RISKS

"To be successful, you have to have your heart in your business, and your business in your heart"

– Thomas Watson Sr.

All projects are subject to risks. Risks have the capability to turn around businesses upside down and vice-versa. Risks can take away all the fun and joy out of the projects or businesses, if not managed well. In the current uncertain conditions, it is difficult for any organisation to sustain and grow without taking risks. However, managing risks is the key to the success of projects. Manging risks can have two main objectives – (a) avoid or minimise the downside of risks, or (b) enhance and exploit upside of risks, called opportunities.

Managing project risks needs an in-depth understanding of various dynamics and complexities (external as well as internal). Well-managed risks can turn out to be an investment, which can yield fruitful dividends. Effective risk management may save projects, businesses, and reputations.

Despite being the highest risky sector, construction is always deprived of good reputation and recognition. It faces continued pressure of managing risks and changes, which adversely affect the predictability of deadlines, budgets with quality deliverables. Unfortunately, there is no simple, single, or straightforward tool or rule to apply and manage project risks. Projects can have known and unknown risks, which may affect during and/or after its project lifecycle.

Thus, project selection is the foremost step in project portfolio management. While the right selection of the project can yield the desired outcomes, the wrong selection can lead to unpleasant outcomes, financial losses, or even loss of goodwill. In such fiercely competitive and rapidly changing times, one of the key risks faced by organisations is the ***selection of the right project.*** Small mistakes in selection can be detrimental to the projects or organisation. Therefore, each project must undergo thorough quality 360-degree assessment – financial and non-financial. Prevalent VUCCA *(vulnerable, uncertain, complex, chaotic and ambiguous)* environment makes the challenge even worse to determine the selection of the right project.

All organisations operate under constraints and limitations. Though any organisation could initiate a large number of projects concurrently to create a magnitude of outcomes, but every organisation has limited bandwidth, resources, and risk appetite. Therefore, project selection needs to be systematic and structured, with a clear goal of a higher probability of success, smooth exit and leveraging benefits as expected or planned. This helps to bring efficiencies for the resources deployed and enable to sustain in the business for a long time.

Preferably, the selection of projects should be carried out by organisation leaders, who have a clear, bigger vision and are responsible to absorb the outcomes. In case leaders are unable to spend adequate time in the project selection process, they must delegate this responsibility to the competent team down the line or subject-matter-specialists, with a clear strategic direction. Leaders should ensure to establish a robust mechanism so that leaders are timely apprised of project selection and right direction or consent be bestowed to the team.

Responsibility of projects or portfolio selection should rest with the leaders or steering committee. The steering committee may comprise of domain specialists, who can add enough value to the process. They should regularly assess, review and monitor project progress to ensure that projects are on-track as planned and in case of deviation or gap, the same be restored immediately. There should not be any disconnect between the leadership and down-line during project execution.

Further, hesitant, erratic, inconsistent, or feeble commitment of leaders in taking project decisions may be a high risk for projects. Such leadership behaviour itself is a risk and may be documented in the risk register. Thus, leadership must be committed to projects, drive it through and own the outcomes. Else, such negligent behaviour or attitude towards projects may become a roadblock to the project performance. Lack of commitment or non-cooperation towards project or ego-conflicts among stakeholders can be serious risk and threat to the project objectives.

Even leaders need relevant information, data, estimates, forecasts, and risks associated with projects to make the right decision. Identification of risks starts from the inception of the opportunity and not after winning the award. Many organisations start risk management process after clinching the deal. Risk culture is to be imbibed from the earliest stage of an opportunity, or before participating in tenders or submitting a proposal to customers.

Usually, project decisions are not made instinctively. The gestation period of project decision is relatively longer than that of the product or transaction businesses. As the stakes are higher in projects, customers may take a long time to prepare techno-commercial feasibility assessment, plans, estimates, budgets, to make effective investment approval and decisions. Thus, complete assessment and negotiations may take a long time to decide.

Risk management should never be the exclusive responsibility of a single individual like the sales manager, tender manager, or project manager. It is a collective and systematic approach to identify the potential risks involved in the project, best done in a group with the help of domain specialists contributing from various internal cross-functional teams as well as external stakeholders. Different methods and techniques can be used to gather inputs from stakeholders and draw inferences thereof.

Evaluation of the following key parameters may support the decision-maker in project selection while participating as contractor or supplier: –

a. **Customer credibility** – Contractor must ensure that the customer is good enough, who can make timely payments. Quality customer is the one who has strong financial muscles and has a good payment history. Current vulnerable market conditions may affect financial conditions of customers anytime. It is not advised to assume that if any customer was financially sound during previous project execution, it shall be the same again. Rather, afresh due diligence should be conducted every time you plan to initiate new project dialogue with a customer – existing or new. It is not limited to its financial creditworthiness but also their non-financial aspects, which include their leadership commitments, internal processes, document approvals, certifications, and payments experiences, etc. In case the customer is not investing their own equity, then the exact sources and status of their funding arrangement should be tracked in advance, so that the project is not jeopardised once activities are commenced, due to payment issues and defaults.

b. **Purpose & application** – Contractor should explicitly understand the purpose and application of the project, including that of its end-user's perspective. It must find out that project deliverables (material or services) are not committed for any proliferation, restricted, illegal, or undisclosed purposes/activities. This is crucial to avoid getting into the zone of unknown risks. At times, the purpose or application of project can impose unknown liabilities. For example, if the material or equipment shall be used in the core production or core processes of nuclear power plants, then contractor or supplier may be obligated under nuclear liability laws where suppliers are unknowingly made liable to nuclear laws. In the absence of purpose and applications, such exposure can be considered as risk of unknown and can bring high exposure for organisations. If unclear, contractor may reach out to end-users or seek written clarifications or indemnifications from end-user. Therefore, it must be confirmed before getting into any legal or contractual relationship with its customer and/or end-users. Wherever needed, the organisation may take advice or consultancies from domain specialists or legal experts or solution architects.

c. **Technical feasibility** – Robust processes must be put in place to thoroughly assess the technical, technological and solution feasibility of the project. Technical team to evaluate the risks involved in technology, design, engineering, solution, capacities, capabilities, makes, standards, codes, and performance testing requirements of the project. While it is good to offer the latest technology or solutions to the customer, at the same time, the contracting organisation must remember that for every new technology or first-time usage, a solution may not work exactly the way it is being sold or told to the customer. Any new product, solution or technology always has its hiccups, features, gaps, or product issues.

However, a contracting organisation should not commit any technology or solution which is neither available in the market (or geography) nor have adequate competency to develop (serve) in required timelines or budgets. Many a time, current seminars, exhibitions introduce advance products or technologies, which are not matured enough to be immediately used in projects. However, their production lines or financial feasibility may not match upfront to consider based on customer timelines and budget constraints. It is also experienced that overcommitments lead to loss of business, cause disputes, financial losses, or loss of goodwill. This may also end up in losing a customer forever.

d. **Financial feasibility** – Cash is king. Unless there are strategic reasons, financial feasibility is the supreme and it supersedes all other factors and reasons in any business. Each organisation must evaluate its commitments, inventories, receivables, working capital and expenditure plans, which can affect its cash positions. Fortunately, there are several financial analytical tools and formulae available to ascertain as to which project may be more beneficial for the organisation. After establishing the technical and operational feasibility, every project must be reviewed based on its financial feasibility considering its return on investment, net present value, internal rate of return, pay-back period, cost-benefit ratios, and cash-flow. The contractor can use back-ups of multiple financial models of costing, cash-flows and other tools. Regardless of the technical and execution feasibilities, if the project is financially not viable, it will be a good decision to abandon or leave the project to prevent losses or misalignment with organisation strategic goals or vision.

e. **Market feasibility** – Irrespective of its alignment with organisation's vision, mission, or strategies, if a project is not aligned with market dynamics, technology, and competition, it is bound to fail miserably or get out of the race quickly.

Competition always drives to reduce costs and improve quality of services, products and deliverables. This can pose risks to aggressive pricing, steep discounts, continued innovation, supply chains and logistics issues and other external economic factors. Sometimes, it may become very difficult for contractors to breakthrough into a new niche, segment or win over new customers, due to competition. There are instances when a contractor may be unaware of their actual competitors during bidding. Whereas at the time of negotiations, they may find that their proposal is being compared or negotiated based on the proposal received from low-tier contenders, which are not at par with OEM or Tier-1 parties. In such cases, contractor ends up losing the job quickly and wasting time, energy and resources. So, knowing the competition, technology, price levels and decision-making criteria of customers is crucial for contractors before participating in any tender or project opportunities.

f. **Execution feasibility** – In spite of the best contract negotiations, if execution feasibility is not properly assessed, then project execution may shatter all dreams and plans. The contractor should assess the location risks, where the project is to be physically executed. Project site may be at an adverse place in terms of its layout, land parcels, land profiles, approach, access etc. where execution or mobility of vehicles, material or manpower is extremely difficult. Such location may have existing difficult pathways, right-of-way issues, hills, forests, deserts, coastal belts, low-lying areas etc. Many projects can pose high risks due to unavailability or unfettered access of full project land, encumbrances and restrictions to land parcels, inadequate workers availability, unusual demands from local people, local mafias, goons, compelling or limited vendors, local laws and other social constraints etc. Local political, social, or regional parties may also be influential in the project location, which can impact project performance, budget and time. A regular supply of construction water, electricity, security and local administration support could be challenging to the projects.

Some tenders, contracts, or framework agreements may require deliveries or execution at multiple locations, geographies, or countries, where exact locations or conditions are not identified or intentionally undisclosed. Such projects may pose the risks of unknown. Any such master agreements, framework agreements or rate contracts may enforce conditions, which may be practically not feasible, as the legal & logistic conditions may radically vary.

g. **Legal & contractual** – While selecting projects, the contractor must assess the tender terms, contract documents, governing laws, regulations, industry practices, associated interventions, and interpretations within a legal framework. Some immature, distressed or desperate contractors have a tendency to accept all tender or contract documents without assessment. This not only includes onerous legal and commercial conditions, but include unfair representations and warranted related to tax structures, exemptions and liabilities, and their futuristic treatments in case of change of law as well. Such negligence or weak assessment could lead to risks for all the contracting parties. The organisation must understand the contract structures, terms and conditions, set of documents, order of precedence, referenced documents and laws etc. which form an integral part of the contract. Inattentive or desperate acceptance of contracts shall lead to a lot of conflicts and disputes between parties and can significantly impact their timely performance.

h. **Strategic alignment.** If you do not know where you are going, any road will take you there. Every organisation has a vision, mission, and strategic objectives. Projects are powerful models to achieve these strategic objectives.

So, while participating in the project selection, leaders or steering committee must validate the project with its strategic alignment. If an organisation has the vision to boom in the renewable sector as part of its strategy, it must stick to its objective and must not divert focus, to remain the long-term winner. Any hoping or unplanned departure from the strategic vision can shift the organisation titanic to some other direction.

In case an organisation is simultaneously evaluating multiple projects and leadership is unable to involve itself rigorously, then a clear selection criteria, limitations, risk appetite levels and delegation of authority must be defined in advance so as to effectively channelise its vision.

Exceptionally, some project financials may not look profitable or reflect good returns in the short terms; however, such projects may be more aligned to the organisation's long-term vision or strategic goals. Therefore, leaders can still make a decision based on its strategic intent or purpose. That is why every organisation needs leaders, as the leaders' thought process is aligned with the long term and not short term.

The picture below reflects how projects contribute to the organisation's Vision and dream.

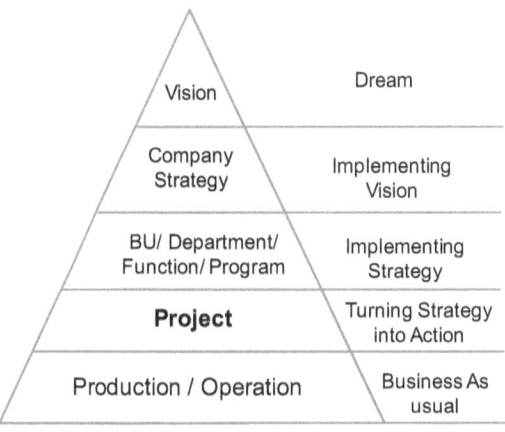

The selection of the projects should fall in line with the vision and strategy of an organisation. It is the responsibility of the business leaders to ensure that projects are properly implementing its strategy, converting dreams into reality and bringing the organisation closer to its vision.

To conclude, it is not only the selection of any project, but the challenge is also choosing the right projects from a wide market and then choosing the project which is expected to give greater results with higher probability, in line with strategic intent. However, the role of the leaders does not end with the selection of the project, but they are also to ensure that it is delivered or executed in the same way as planned. It means regular review and monitoring of the project are equally crucial throughout its lifecycle.

Chapter 3

PARTNER SELECTION RISKS

"Individually, we are one drop. Together, we are an ocean"

– Ryunosuke Satoro

Selection of partner(s) is critical for any project. The success of a project depends heavily on the capacities and experiences of your partners – be it customer, owner, contractor, supplier, consultant, consortium partner or other business associates. Selection of an incapable partner may jeopardise the project itself. Therefore, partnering is a mutual commitment and investment of parties, where time, money, resources, experiences, and risks are exchanged for common goals and objectives. This makes it essential to conduct a careful assessment of the partner, which can otherwise affect the project seriously.

Due consideration must be given while selecting partners: –

- As a customer – If you are appointing a contractor, supplier, consultant, or service provider, make sure that you appoint a capable party, who can ***deliver on time!***

- As a contractor – If you are willing to enter into a contract with a customer or buyer, ensure that the customer has adequate financial muscles, who can *pay on time!*

- As a consortium – If you are willing to take up the project in a collective partnership or consortium arrangement, select partners, who can *stay committed long-time!*

It is not only the financial or commercial strengths of the partners, but their non-financial credentials are also equally important like technical competencies, operational capabilities, relevant experiences, core values, and strategic alignment. Many organisations accord high priority to ethics, quality, safety, compliances, and reliability of its partners so that long-term relationship and goodwill are established.

1. Selection of Customer:-

When you are a contractor, supplier, consultant, or service provider, the most important risk to mitigate is the commercial or financial aspects related to the customer. Your customer must have adequate financial muscles to make full timely payments as per the terms and manner agreed in the contract.

Some levers that a contractor can adapt, to secure its payments include:-

1.1. **Status of the legal entity** – Many personnel get quickly influenced or excited with the name of a large cap company group assuming it is a good customer account, whereas it is true that bigger is not necessary be always better. To safeguard commercial interests, one must find out the exact name and details of the legal entity, which is going to enter into the contractual relationship. Many customers initiate the procurement dialogues in the name of its parent or group company, which has a renowned presence but while awarding the orders, they switch to other subsidiaries or special purpose vehicles (SPVs) company. Contractors should seek its registration evidences, including shareholding, taxes and exemption-related qualifications. These subsidiaries or SPVs may not have equal or adequate financial strengths and/or past financial records to establish creditworthiness.

Most green-field or large infrastructure projects are set up in the name of new SPVs or independent construction companies. Most customers, developers or builders float new special purpose vehicle (SPV) company or joint venture (JV) company for the construction of a new project, which may not have necessary approvals and credentials at the beginning. It may be difficult to get the credit insurance for such SPVs or JVs, in the absence of past credentials or records. As a result, the contractor needs to find an alternate mechanism to protect their interests through bank guarantees or corporate guarantees from the customer's parent or flagship entities. However, such instruments can be invoked in the case of failure of the contracting company to honour its commitments. Therefore, risks associated with legal entities must be carefully assessed, analysed and mitigated before getting into contractual engagements.

1.2. **Customer creditworthiness** – Contractor should evaluate the creditworthiness of the prospective customer with the help of financial experts. The creditworthiness of a company can change at any time and do not assume the same forever. If the parent or flagship company of a customer has strong creditworthiness, it does not necessitate that its subsidiaries or SPVs shall also enjoy the equal or similar benefits or credits in the market. Hence, one must thoroughly assess the creditworthiness. Frontline team of an organisation may insist to follow or assume the previous or old creditworthiness of a customer. This can turn out to be a high risk by not verifying the latest status of the customer. If a contractor has executed projects in the past for a customer or is executing projects in a staggered manner, still it is advisable to check the status of current creditworthiness of the customer. Any such statement of any employee that "I know my customer well" or "This customer has always paid us on time in the past" or "None of my customers defaulted in the past" or "I have strong relations with the customer," etc., all such statements or emotions do not hold water in case differences, conflicts, disputes or litigations arise. So, it is strongly recommended that while assessing the customer risks, no such personal biases be attached during verification.

1.3. **Payment experiences** – If an organisation has dealt previously or is still dealing with a customer, then you must have adequate experiences of their payment behaviour. Many customers may look easy to deal with but factually their internal processes are so complicated, or portfolio priorities are different that may affect the timely release of payments. Thus, you must leverage the past payment experiences of such customers. If you have not dealt previously, then adequate attempts are to be made to find out from market about the trends of customer payment to its other contractors and suppliers and average DPO (days of payment outstanding). Many contractors map the complete invoicing and payment collection process in the beginning so that a realistic prediction can be made regarding cash-flow. Many customers follow lengthy certification or documentation process before payments. Even the verification of multiple documents from construction sites or sharing the documents at different locations may add complexity to payment collection. Many customers may not agree for interest charges on delayed payment, thereby affecting project cash-flow. Contractors can also explore market commensurate interest charges for late payments or vendor discounting options for early payment realisations.

1.4. **Source of funding** – As each project is unique, exclusive and sources of funds may be different. Customer may infuse self-equity or borrow from other financial institutions or lenders. It is safe to know the exact source of funds for the customer, so that contractor payments are not stuck and the contractor can assess the associated risks in advance. If the customer is borrowing from external sources, the contractor should find out the exact status of their funding, financial closure, and disbursement plans. Generally, financial institutions have complex processes and require a large number of documents which take a long time for customers to convince lenders. Thus, it may take a longer time to get the loan sanctioned from financial institutions. Sanction of loan does not mean the disbursement of the loans. To protect the contractor's interest, financial closure may be a 'condition precedence' to Notice to Proceed (NTP) of the contract. It is worth remembering that many organisations may not commit high costs before financial closures or firm commitments are received by them.

1.5. **Payment mechanism** – Payment security is key financial risks for all businesses. Various payment modes are used in the construction market. It may include cash payments, wire transfers, documentary credit, letter of credit (LC), vendor financing, etc. Ideally, full advances before shipments or services or opening of letter of credit payments are assumed the safe mode to secure payment from a contractor standpoint. All international trade transactions take place with prior opening of confirmed, irrevocable and unconditional letter of credit through mutually agreed banks. However, for domestic contracts, the customer provides initial part advance payments through cash or bank transfers. Balance payments are made progressively by the agreed mechanism, upon achievement of the agreed milestones. Most contractors insist on payment security through letter of credit. In order to protect payments risks, contractors should seek payments through LC and it should be opened along with manufacturing clearance or before the start of manufacturing so that upon readiness of the material, shipments are not unnecessary withheld and contractor payments are not affected. These days, invoice or bill discounting or vendor financing models are also very popular wherein contractors or suppliers can get the payments in advance at discounted rates from the agreed lenders or banks.

1.6. **Terms of payments** – Delays and discrepancies at the time of payment is a very common deterrent in projects. Generally, project payment terms are "open account" (supplier provides the material or services to the customer without any guarantee and expect payments at later date) through the exchange of certain documents. Although, advance receipts of cash is the most secured payment terms. Timely payment receipts should be the essence of the contract. Payment terms are the most critical element in setting the invoicing and collections clearly, which directly affect the project cash-flow. As contractor, one must negotiate for advance from its customer. Common industry practice is 10% to 20% of the total contract price, provided as down payment to the contractors (with or without advance bank guarantees of equivalent amount). Balance payment terms are made progressively upon deliveries or performance of work. Credit period and document requirements must explicitly be agreed in the payment terms. In turnkey projects, customers may hold retention amount (5% to 20%) and release it after the achievement of agreed milestones (like commissioning, substantial completion, provisional or final acceptances or hand-over etc.). Thus, the retention criteria and requirement for release of withheld payments should also be explicitly agreed in the contract itself. Many customers demand performance bank guarantee, as a token of commitment towards its warranty or defect liability period obligations. Contractors can negotiate for early release of cash against bank guarantees so that cash is used in business. It is also strongly advisable that while negotiating the payment terms with customers, the contractors should also know the payment terms with its down-line vendors and sub-contractors, whereby back-to-back better terms are agreed.

1.7. **Bank guarantees** – Key instruments that safeguard customer interest is the bank guarantees or security bonds issued from authorised banks or financial institutions. A critical point to remember is that *most bank guarantees provided to customers are usually unconditional,* and the customer has full rights and privileges to invoke or encash the same, (without seeking prior consent from the contractor), unless otherwise specifically consented in the agreements. Down payments are released by customers against submission of advance bank guarantee (ABG) and/or contract performance bank guarantees (CPBG). Value of ABGs can be equivalent to the corresponding amount being released and valid upto project execution period. Further, many contracts require its contractors to submit performance bank guarantees (PBG) or warranty bank guarantee (WBG) before the release of retention or final payments. These PBGs or WBGs are provided for an agreed period, as a measure to demonstrate contractors commitments to discharge its contractual obligation during agreed (warranty or defect liability) period. In case the contractor fails to rectify or repair defects, damages, or performance issues during the agreed period, the customer may encash the BGs and get the work done on its own. This is the alternate recourse available with customers, besides legal remedies provided under the law. In the event, contract BG is required for a longer period (5 years or more), then contractors may negotiate the rolling bank guarantees, which can be renewed year-on-year basis as well. Commonly, the formats of the BGs are overlooked by contractors or suppliers, which becomes a point of concern while issuing the same. Thus, a thorough review and assessment of the BG formats must be done in advance, as it also forms an integral part of the contract itself.

1.8. **Payment documentation** – Every customer mandates certain documents to qualify contractor payments. These documents are required to establish the genuineness, authenticity, correctness, and completeness of the transaction for which payment is due. Thus, the contractor should carefully agree on these documents, to be furnished along with each payment demand.

Advance related – Most contracts demand issuance of advance payment at different stages like (a) upon execution of the contract, (b) site mobilisation, (c) submission of preliminary design/drawings or (d) award of vendor or sub-contractors orders etc. Parties should clearly agree on the document requirements and tax considerations, which required for release advance payment.

Material related – To prove as evidence of material inspections, dispatches or delivery, the contractors may require to submit various documents, which may include – tax/commercial invoices, stage or factory acceptance tests, inspections reports, dispatch authorisations (DA), material dispatch clearance certificates (MDCC), packing lists, delivery challans, lorry receipts, bill of lading, way-bills, etc. Some documents may also require certification (sign/stamp) from customer site stores or acceptances by their authorised engineers.

Service-related – To prove as the evidence of services rendered (like design, consultancy, civil, mechanical, electrical, plumbing, sanitary, erection, inspection, testing, commissioning etc.), the contractors may require to submit various documents may include – invoices or running-account bills, certification of works, measurement books, logbooks, workers wage registers, labour compliances, material reconciliation document etc.

Many customers may demand multiple set of documents, to be delivered at different locations. So, such document requirements should be assessed thoroughly to avoid unnecessary discrepancies and delays in payment realisation. Document related issues, discrepancies and disputes are very common, which becomes a bottleneck for release of payment, despite complete the physical obligations by contractors.

2. Selection of Contractor (Supplier):-

When you are buyer or owner or customer, you may engage contractors, suppliers, consultants, and other service providers to carry out the various activities. While selecting down-line partners, the highest risk you pose is their technical and operational capability to deliver the project on time. Potential contractors or suppliers should have adequate technical know-how, capabilities and competency to depute resources and matured processes to meet the project objectives within constraints.

Some levers, customers can adapt in contractor selection, to protect its interest includes:

2.1. **Contractor's creditworthiness** – Customer should evaluate the creditworthiness of the potential contractors or suppliers with the help of experts. Each contractor should be capable of: (a) timely hire or deputation of competent manpower, (b) deployment of required construction material – tools, tackles, machines, etc. (c) purchase of raw material for manufacturing or construction works to carry out activities, (d) timely payments to its employees and vendors and discharge its statutory obligations. Initially, the contractor may need to invest some money to arrange resources and sustain, until a cash-flow cycle comes into rotation, though some contracts have provision for advance payments to contractors. However, many contractors may not have adequate resources, bandwidth or even bank limits to provide a bank guarantee itself or procure the initial raw material. As a result, this becomes a huge open risk for the customers, as such contractors can also threat, stop or reduce the pace of work. Hence, one must evaluate the creditworthiness of the potential partners. At times, internal stakeholders may influence to depute specific partners and assume their past records or credits. This can turn out to be a high risk by not verifying their creditworthiness and the party can turn out to be bankrupt or withdraw resources due to non-feasibility during project performance. Any such statement of any stakeholder that "I know this supplier" or "This contractor has always executed on time in the past" or "None of my contractors have defaulted in the past," etc., all such statement hold no water in case of conflicts, disputes, or litigations. So, if you engage any cash-strapped partner, you are opening a big Pandora's-box for big issues during project performance, who can grind or arm-twist the customer upon project commencement.

2.2. **Capacity to produce** – Most large projects usually include the supply of material in the scope of contractors. In such cases, parties must agree on makes, models, manufacturers and origin of every material, besides specifications. Accordingly, parties can ascertain the manufacturing capacities of selected vendors and sources. Manufacturer's capability becomes a big bottleneck when critical items require customisation as per specification or involve a high volume of material. Sometimes, a small vendor may receive multiple orders simultaneously and fail to allocate full capacity to a single customer or prioritise their supply well for a specific project, as such vendors may divert their supplies to other projects of their choice or interests and can jeopardise the project. For example, recently, the Government of India has given high impetus to the development of a large number of renewable projects and associated grid/transmission networks. All renewable plants and transmission networks have a high demand for structural material for module mounting, switchyards, grids and transmission systems, whereas good vendors for structural material are limited or have capacity constraints. In case these suppliers receive high-volume-orders from multiple customers simultaneously, these vendors may fail to execute all the orders concurrently and may impact the supply plans for most customers, which can have a direct impact on project schedules. Hence, proper assessment on their manufacturing capacity, existing order loads, sources of raw material, manufacturing processes and allocation of capacity must be done carefully to de-risk your project supplies.

2.3. **Quality requirements** – During pre-award or order negotiation stage, customer and contractor must discuss, agree and sign off the key aspects of manufacturing quality plans, field quality plans, quality standards, quality procedures and selection/rejection criteria and relevant checklists and formats, etc. Once this is converged between the parties, the same must form part of the contract. This shall ensure that there is no gap between the parties and the project is executed smoothly. During the selection of the contractor, the customer should visit their manufacturing units or other project site locations to witness the actual working processes, standards, compliance levels, cultures and all other quality aspects of suppliers' capabilities. Customer should also demand or verify the documents related to valid type tests, certifications, standards, accolades, and quality performance credentials acquired by contractor or supplier, as rejection of any critical material or field activity or work can significantly affect the overall project schedule, lead to conflicts between the parties and jeopardize the project. Parties should expressly agree upon the category of material and mechanism for its inspection, packaging, dispatch, acceptances and certifications by authorised persons or agencies.

2.4. **Health, safety & environment** – Safety of humans, property, environment, and society are common risks in construction. These are not only legal or statutory requirements but also big social responsibility for corporates. Customer should clearly prescribe its requirements and expectations at the beginning of the project and include the same in the contracts. Some customers may require a specific set of PPEs, tools, resources, lab-setup, labour setups, medical staff/facilities, tie-ups or arrangements with specific agencies or procedures to be adhered to while executing the projects. Commonly, contractors do not consider adequate budget for such requirements, which can cause unreasonable stress or differences between parties. This places not only the contract performance at risk but also the goodwill of the companies. Hence, this risk must be mitigated with the help of HSE specialist and HSE plan is agreed and put in place.

3. Selection of Consortium Partner (As Contractor):-

Many contractors may not participate in tenders or project opportunities alone and explore an alternate route of consortium arrangements along with other partner(s). Such arrangements can be made to use other's credentials, technical know-how, financial, operational, distribution of risks or other compelling reasons which instigate them to join hands with other partner(s). Consortium arrangements bring a different set of risks and challenges for the parties or even the project itself.

Consortium arrangement may be explored for various reasons:

- **Qualification** – A party is individually unable to meet full preliminary qualification requirements (QR) of tenders or projects.
- **Capabilities** – A party is individually unable to fulfil or meet complete project scope, solutions, deliverables or requirements.
- **Technical or operational feasibility** – A party is individually unable to accept or execute the complete technical or operational scope of large projects.
- **Risk appetite** – A party is individually unable to invest heavily on the project or take fully financial risks. Therefore, risk can be distributed between partners.
- **Strategic reasons** – A party individually may not have experience or internal competencies to fulfil project requirements or is venturing into a new field. Such credentials or learning can be leveraged by partnering with other companies.

Selection of a weak or incompetent partner can severely affect the project performance. While deciding the consortium partners, the contractor should evaluate the following key factors before entering into a contractual agreement with its main customer: –

1. **Number of partners:** It is advised to restrict the total number of consortium partners to essentially minimum. Unless inevitable, consortium should have maximum two-three parties. The more the parties, the more the chances of conflicts, confusions and complexities for all partners.

2. **Party liabilities:** Parties should carefully agree on liability models. Many customers or end-user may enumerate specific liability models. So, parties must evaluate and adapt the liability models. Following models are commonly used:-

   2.1. All contract parties (consortium partners) are *'Jointly and Severally Liable'* for the total project scope and value, regardless of the individual contribution of each party. It means, even if a party is contributing to only 10% of the total scope work, all parties become legally liable for 100% of the total project scope of the full contract, which is 10 times higher than its individual scope of work.

   2.2. Leading party or primary partner is *'Jointly and Severally Liable'* for the total project scope and value and other partners be severally liable to the extent of their respective scope of work or contract value of their scope only. It means, even if the leader or primary partner is contributing only 10% of the total project scope, it still becomes legally liable for 100% of the total project scope, which is 10 times higher than its individual scope, whereas, all other parties are liable for their respective scope of work only. Therefore, such a position should be taken only when the leading party has the highest scope or contribution in the total consortium arrangement.

   2.3. All contracting parties or partners are '**Severally liable**' to the extent of their individual project scope of work or contract value. It means, the default or delay of any party's obligations shall not affect other partners. In such scenarios, although multiple parties are collaborating through a consortium, they are absolved from the risk, responsibilities and liabilities of other's scopes or obligations.

Unless there are compelling reasons, it is preferred that the party who has the maximum share in terms of scope or value should be the leader of the consortium.

Besides the selection of above models, consortium partners should thoroughly evaluate the processes and consequences, in case of default, termination or back out of any party after the execution of contracts. This becomes all the more challenging when contract periods are long or/and complex scopes are involved.

3. **Division of responsibility (DOR):** Explicit distribution or division of responsibility helps to establish clear accountability. Before executing a consortium agreement, all parties must clearly define and converge on their respective scope of work (SOW), obligations, warranties, performances, responsibilities, liabilities, payments, penalties and consequences. Parties must have written sign-off documents with proper DOR & SOW and incorporate the same in the contract, to de-risk conflicts or ambiguity during project performance.

4. **Schedules & inter-dependencies:** – All consortium parties can work concurrently in a project for their respective scope of work. However, it is not necessary that all parties shall mandatorily be engaged for the full tenure of the project. Any party can start early or late for their respective scope of work. However, in case there are inter-dependencies, the integrated project schedule must be agreed with the intricacies and dependencies of the parties, as delay of one party may impact another party and/or the overall project. So clear schedule and communication mechanism be established between the parties. Any delays or dependencies should be timely brought up during progress reviews and action plans to overcome challenges shall be evolved.

5. **Financial mechanism** – Consortium parties must agree on payment mechanisms at the start of the project. Some customers restrict that all invoices, payments, reconciliation, penalties, and adjustments are dealt only with the leader or primary party. In such cases, the leader or primary partner should align or fix the '*back-to-back*' or '*pay-if-and-when-got*' terms basis arrangements with its associated partners. This may also include seeking back-up guarantees and warranties from other partners.

6. **Securities & guarantees** – Many contracts require all consortium parties to submit the required earnest money deposits (EMD), securities, bonds or guarantees to the main customer for their respective SOW. However, if the customer restricts that such instruments can be submitted by the lead bidder only, then back-up arrangements between/among the consortium partners must be done separately, wherein all other partners can submit the proportional EMD, securities, bonds or guarantees of their corresponding value to the primary leader. In case of forfeiture of such instruments by the customer, the treatment of other partners' instruments must be considered carefully, as forfeiture by the main customer may or may not be attributed to other parties and what if it is exclusively the default of the leader or any specific partner.

To conclude, the selection of your partner is crucial for the success of the project. So, an effective congruence should be assessed for partners related to strategies, structures, skills, capabilities, processes, and shared values. Depending upon the requirement and positions of the partners, a structured process must be evolved thereby involving relevant subject matter experts from cross-functions and enabling leadership to decide the most suitable partner.

Chapter 4

FINANCIAL & COMMERCIAL RISKS

A small profit is better than a big loss

– Ron Rash

Financial and commercial risks can increase costs, reduce sales or profitability from what was planned or forecasted. The current project environment is very dynamic, fuzzy, risky and uncertain. Despite deploying best project specialists and working out the best estimates, there are several factors, which remain in a situation of flurry.

Financial risks may emerge due to:

- *Instable political conditions*
- *fluctuating market and economic conditions*
- *Scope changes or scope creeps*
- *Incorrect assumptions or errors*
- *Poor processes and controls*
- *Cost of prolongation*
- *Change in law*
- *Many other external factors, etc.*

These vacillating factors may include a surge in commodity prices, fluctuating foreign currency rates, interest and credit rates, market demands, payment delays, unmanaged cash-flows, inaccurate estimations or assumptions and frequent changes in scopes and requirements, etc. Most of these factors pose direct financial risks for the projects.

All other risks highlighted elsewhere in this book may have a direct or indirect commercial impact. Unless risks are managed properly, it may severely hit the project financials. However, early detection of risks and proper risk management may reduce the damage or may even turn a risk into an opportunity so as to achieve or improve forecasted numbers.

Cost overruns and delayed completion in the project are common phenomena in construction projects and directly impact the project performance. Managing projects' budgets and cash-flow, are major risks, which keep the stakeholder on toes throughout the project lifecycle. Many organisations strongly believe that their projects are driven by numbers alone, as rest of the backward actions follow the targeted numbers. Thus most business leaders closely chase committed numbers and deviations in reviews or assessments.

Cash is king and running out of cash is often the endpoint in projects or businesses. Some key questions which every project person must ask:

- Where does the cash come from?
- Do we get paid as committed?
- What can hit cash—unexpectedly and uncontrollably?
- Can customer become defaulter or bankrupt?
- Can market and economic dynamics impact the project?
- How changes (internal or external) can hit project?

Budgets are the backbone of projects. Projects operate in uncertainty and vulnerability, which bring high risks for projects. In such situations, if contractors commit the certainty, then customer finds value in it. Thus, most large projects are preferred on a turnkey basis with '**Firm Fixed-Price**' (FFP) model. Under such contracting arrangements, the customer agrees to pay fixed prices for a defined scope, which are not subject to any fluctuations or adjustment based on contractor's costs in performing their obligations. Such contracts place the maximum risks, responsibilities and liabilities for costs (with resultant profit or loss) on contractors' account. In case the actual costs of execution exceed the estimation or assumption of the contractor, the contractor has to suffer the consequence of financial loss or goodwill. Therefore, FFP contracts relatively cost higher. Thus, such contracts are becoming popular as compared to variable contracts which give less control to the customers. These contracts boost customer's immunity over the uncertainty of volatile markets and conditions.

FFP contracts pose high risks for contractors as compared to variable, cost reimbursement or time and material contracts. The contractor should make complete and precise estimations, for all costs, overheads, profit margins and contingencies before arriving at the final price. If any item, task or activity is missed or under-estimated, the projects may face financial jolts and jerks.

In the current competitive world, most projects face acute pressure of excess costs and cutting costs from all nooks and corners, although, companies may make aggressive decisions of cost optimisation by compromising timelines, qualities, resources, features or other parameters. This can also impact by shrinking the planned profitability. Contractor's return on investment has a high dependency upon risks in the project. Hence, a systematic and robust system should be adopted to ensure that projects deliver the committed returns on completion.

Obtaining valid back-up commitments from vendors and sub-contractors, reasonable estimation of timelines (with realistic prolongation), clarity of scopes and obligations are keys to ensure improved chances of success and avoid undue losses or stresses later.

Needless to state that many organisations may weigh various options related to designs, technologies, supply chain, contract structures and execution methodologies to optimise costs, provoke competition or de-risk dependencies on monopolistic party or parties.

Some of the financial deterrents and risks in projects include:

1. Funding Risks

Due diligence of customer's current financial health, past track record, sources and status of funds, etc. is the key to ensure the financial stability of the customer. Certain contractors succeed to interweave the contractual binding commitments from their customers about the availability of funds and payments. As absence or delayed availability of funds or budget allocations increases the risk of failures or consequences in meeting its goals. So, it is advantageous to seek absolute clarity on the following aspects and assess the associated risks:-

- Understand the types, sources and status of funds and funding mechanism
- Impact of failures to secure funds or refusal of lenders for project funding
- Status of customer liquidation or bankruptcy

It is also worth evaluating the fund statuses and mechanisms adopted by the customer in past projects or other ongoing projects.

Payment security through letter of credit, financial guarantee from customers, or parent/group guarantee in case of customer default are some of the good measures to mitigate the funding risks. These are common and high impact risks which can cause direct significant impact on projects. Many customers may explore construction-linked funding, which is disbursed as the project progresses.

2. Revenue Credit Risks

In view of the present volatile situations, it is good practice that contractors may take revenue credit insurance against a nominal premium, so that if the customer fails to make the due payments or goes bankrupt, the revenue credit insurance company can settle the cost to a large extent and save contractors from bigger jolts and jerks. There are many insurance companies which can offer revenue credit insurance based on the credentials of the customers and risks involved in the project. Contractors can also mitigate such risks through revenue credit insurance covers.

3. Project Costing Risks

Project performance and quality of deliverables largely depend upon the project costing and cash-flow. If the actual cost exceeds the planned estimates, then the forecasted profitability may be impacted. This may force the contractor to reduce the scope or compromise on quality to stay within the baseline budget. Costing risk may also lead to schedule risk if the schedule gets extended due to non-availability of enough funds to accomplish the project scope. Thus, systematic and balanced control is required to manage the project within triple constraints—budget, time, and scope. Else these constraints shall compound the risks in the projects.

There is an irony that even if everything else goes right in a project, but cost estimates are defective, then it affects the project severely. Thus, accurate financial estimates are crucial for project success. To make correct estimates, the team should have good understanding and clarity of scope, requirements, commitments and expectations, market costs, forecasted price movements, past learnings and associated risks. Moreover, errors, omissions or issues in the costing may not be conspicuous or detected immediately. Yet the consequences of defective costing could be devastating for the project. Correct, complete and accurate costing is the key *mantra* to de-risk financial exposure. Another key risk is the timing of incurring the costs, as good supply chain experts should know the trends of commodities and market speculations before deciding the timing of orders on vendors and sub-contractors.

For complex projects, if proposal managers, who may not have a finance background prepare the project costing, it could be a risk, as proposal managers may be good in understanding project scope and technical aspects but may not do justice to financials. Likewise, finance persons may be good with financials but may lack the understanding of the scope, technical aspects and visualisation of site situations as considered in costing. Hence, the collaboration of technical and commercial experts is important for good costing and minimises the financial risks in the costing.

Mostly, costing personnel work under target-pressure to meet customer targets, market levels and to appease bosses, thereby compromising on many fronts in costing, which becomes a major deterrent during the execution phase. For example, proposal managers may consider the lowest cost of key material from approved makes, as per Bill of Material during cost estimation, whereas when actual orders are placed, the lowest cost vendor may not meet the delivery timelines, exact specifications or asks upward price due to commodity inflation; then companies are bound to take a financial hit by placing orders on another vendor with a cost higher than estimated.

Project cost estimation can be affected due to various reasons:

a. Technical experts lack financial knowledge and may lead to incorrect estimates.

b. Finance experts lack understanding of project scope, technical complexities, site data, site conditions and constraints, and other approvals and permissions required for the project. Thus, they become a sheer consolidator of inputs and fail to add true value.

c. High dependency on standard costing templates (in MS Excel or other software tools) to solve all costing issues and problems. Finance persons must ensure that all computations are correct and there are no errors, omissions or mismatches in the formulae.

d. A dedicated expert is not assigned from the beginning, who can track all material, resources, overheads, profit margin, changes and ensure correct costing. This person can ensure to consider final back-up offers, eliminate redundant costs and avoiding unnecessary mistakes and conflicts.

e. Limited resources or overloaded tendering teams to prepare a large number of bids concurrently, which not only increases the probability of errors and mistakes but also cause to miss the opportunity of proper technical-commercial evaluation. Thus, all proposals should be closely validated by different technical and commercials members, without compromising the confidentiality.

f. Many organisations consider a lump-sum contingency in costing rather than have proper risk assessments to ascertain more effective risk cost. This leads to incorrect assumptions of risks on fluctuations, inflations and other financial costs.

g. Disregarding the lessons learnt from past projects. Every project gives a lot of lessons which impact financials positively or negatively. Each time a project is completed, it is a great opportunity to reflect or revisit cost estimations and imbibe the lessons for future project costing. If these are not incorporated in your costing templates or formulae, you will repeat the same errors in future projects as well.

h. Incorrect categorisation of the costs. Organisations may split the costs into major categories like (i) Fixed and variable costs; (ii) Direct or indirect; (iii) Billable or non-billable costs; (iv) Product and period costs; (v) Project or corporate costs, etc.

Overcommitments can have major dents on cost estimates. Thus, the team should make realistic assumptions and commitments. Before making any additional or impulsive commitment to customers, assessment of its impact should be duly evaluated to safeguard project financials. Project costing has no line item to absorb the impact of relationships or emotions.

Projects are unique and never go 100% as planned. Therefore, the project team has to anticipate what might happen throughout the project lifecycle (good or bad) and give realistic estimated timelines. There is a tendency to make over-optimistic plans, watertight schedules and aggressive resource planning, which are not achievable during real execution. Ultimately, they lead to time and cost overruns.

Considering last purchase prices as base references for costing can have its own consequences. As prices of past purchase orders may not be valid anymore, fresh orders may be placed after a gap of a few weeks or months. Input costs are significantly impacted due to economy, commodity and currency volatility and changes.

If projects are delivered on-budget, on-schedule, on-scope and on-quality, the outcome shall be fruitful and performance is appreciated. But in VUCCA environment, there is high volatility and probability of changes. Every cost-overrun can impact the bottom-line of the project. Many times, contractors make decisions in a hasty manner based on thumb rules, which may lead to expensive mistakes and undue stresses.

Some components compromised in project costing lead to risks, failures or losses:-

a. Incomplete understanding of project scope, deliverables and requirement

b. Unclear 'makes,' 'vendor lists' or 'origins' of equipment and material

c. High exposure from material issued by customer (free issues)

d. Under-estimated scope of services or non-BBU items (engineering, packaging, logistics, services, operations and other miscellaneous items)

e. Financial loading – Incorrect estimations of interest rates, delayed payments, hedging cost, warranty or evergreen warranty costs, storage and inventory carrying costs, insurance premiums and overheads, etc.

f. Excessive fluctuations in commodity or currency

g. Taxes, duties and other levies, (including change in laws and applicable exemptions)

h. Project extension further withheld retentions, extension of bonds and guarantees

i. Unrealistic site setup costs and weak resources planning

j. Incorrect assumption of time, costs and efforts for regulatory approvals and permits.

k. Cost overruns due to prolongations of execution or closures.

A common error in costing is discovered about missing to consider the revised wages and salaries of manpower year-on-year and third-party commission payments. The salary and wages of the team may get revised every year, but costing does not reflect or cover the increased wages or bonuses, including their extended stay if the projects are delayed.

Most projects take longer time than envisaged during closure, which impacts heavily on planned budgets due to extended stay or financial costs, which are not considered during initial costing stage to become competitive or out of negligence.

4. Project Cash-flow Risks

With regards to cash-flow, every project is to be treated as an independent venture. It is prudent that every project should be able to manage its cash requirement out of its own cash inflow and maintain a positive cash-flow. Positive cash will help to optimise the funding cost which will be instrumental to increase profitability and competitiveness.

Due assessment of customer's creditworthiness (background, financial health, payment experiences, DPOs, market or creditworthiness, source of funding, other ongoing investment of projects etc.) supports to establish whether the customer will be able to make the payment as per agreed payment terms. This is the first lever of assessment to ensure the positive cash-flow in its entire life cycle.

Critical measures to be adopted by the contractor to ensure positive cash-flow like:-

- Clearly define payment terms, credit periods, and billing break-ups
- Complete billing process and documents to be annexed thereof

- Ask for maximum advance payment with order placement or mobilisation

- Progressive payment milestones like –
  - Documentation – Agree on layouts, drawings, specification, or other documents, to quality the payments and deliverables, at every stage
  - Supply-related – Agree on supply-related units, milestone of material, equipment, Incoterms of material, for transfer of risk and cost
  - Service-related – Agree on units, milestones, inspections, completion, acceptance, and handing-over processes
  - Retention – Amount retained progressively from each bill and claim mechanism

- Back-to-back or better payment terms with main suppliers or sub-contractors

- Negotiate longer credit period with the suppliers or sub-contractor, as compared to customer's credit periods

- Negotiate performance bonds and securities (in lieu of cash retention) with customer and receive back-to-back from down-line suppliers or sub-contractors.

Forecasting made during planning or bidding stage varies in actual execution, due to multiple reasons. Thus, it is essential to closely monitor the cash-flow estimates of planning versus the actual cash-flow achieved during execution. This shall significantly help companies to take immediate remedial (corrective or preventive) actions to bridge the gap so that the projected cash-flow target is achieved, or negative impact is reduced. This regular exercise must be carried out from the beginning. Rather, organisations may assign dedicated roles or teams to manage cash-flow and payment collections. Bill-generation, submission, processing and cash realisation undergoes a large number of hands and approvers, which is a grey area in most organisations. A proper monitor or expediter can contribute significantly to cash management. It should be categorically understood that billing and collections are two independent processes and should be mapped and tracked separately.

5. Payment Risks

Timely receipt of full due payments is uncommon in the construction industry. So, unless contractors are doubly sure of their payments from the customer, risk of delayed payments must be properly documented in the risk register.

Due to onerous payment terms, customer's complex processes to pay, contractor's failure to negotiate better or back-to-back payment terms with its vendors, etc. can hit the cash-flow. So, the contractor should make a serious attempt to negotiate good payment terms to manage cash-flow. However, many hurdles, situations or discrepancies during execution can also contribute to turning the cash-flow into the red or leading to payment disputes.

In case the project is having a negative cash-flow, the finance cost at the prevailing interest rate for borrowing to the extent of negative cash-flow, is to be charged to the project to ensure accuracy in project costing. It is considered that the project may receive short-term funding to manage the cash out of obligation and interest paid for the funding.

6. Case Study: Cash-flow

M/s. XYZ Company received a prestigious order of 1000 MINR for construction of switchyard along with associated distribution network, to be delivered in 12 months. During the construction phase, the project had to undergo multiple change orders, which led to slippages of timely work certifications and payment receipts. Inevitably, XYZ Company adjusted such changes in cash-flow, from what was envisaged during the initial bid submission stage.

As the end result of the changes, the project value was revised to 1120 MINR (with positive change orders of 120 MINR). Project timelines were revised to 14 months (with an extension of 2 months). Project gross margin anticipated as 30% at the time of award of the order was revised at 28% at the project closure. Comparative financial and cash-flow statements at the two stages are given below:

Cash Flow-Tendering

Outflow

Details	M1	M2	M3	M4	M5	M6	M7	M8	M9	M10	M11	M12	Total
Material	-	-	-	22.5	45.0	90.0	90.0	45.0	67.5	67.5	22.5		450.0
Project Mgt	2.0	2.0	2.0	3.0	3.0	3.0	3.0	3.0	3.0	3.0	2.0	1.0	30.0
Design & Engg	2.0	2.0	3.0	6.0	6.0	4.0	2.0	2.0	1.0	1.0	1.0		30.0
Services						6.0	8.0	10.0	12.0	10.0	9.0	5.0	60.0
Risk %												40.0	40.0
Warranty												40.0	40.0
Others	4.0	4.0	4.0	4.0	4.0	4.0	4.0	4.5	4.5	4.5	4.5	4.0	50.0
Total	8.0	8.0	9.0	35.5	58.0	107.0	107.0	64.5	88.0	86.0	39.0	90.0	700.0

Inflow

Details	M1	M2	M3	M4	M5	M6	M7	M8	M9	M10	M11	M12	Total
Adv. 5% (Contract)	50.0												50.0
Adv 5% (Drawing)		50.0											50.0
Supply - Progressive			40.0	80.0	160.0	160.0	80.0	120.0	120.0	40.0			800.0
Services											100.0	100.0
Total	50.0	50.0	40.0	80.0	160.0	160.0	80.0	120.0	120.0	40.0	-	100.0	1,000.0
Cash Flow	42.0	84.0	115.0	159.5	261.5	314.5	287.5	343.0	375.0	329.0	290.0	300.0	

Cash-flow prepared and estimated by the tendering manager during bidding stage was completely positive throughout its lifecycle. But upon start-up of the project, when the project undergoes multiple changes, the cash-flow is also impacted as under.

Cash Flow - Execution

Outflow

Details	M1	M2	M3	M4	M5	M6	M7	M8	M9	M10	M11	M12	M13	M14	Total
Material Cost	-	-	-	27.5	54.9	109.8	109.8	43.9	82.4	82.4	27.5	11.0			549.0
Project Mgt.	2.0	2.0	2.0	3.0	4.0	4.0	4.0	4.0	4.0	5.0	3.0	3.0	2.0	1.0	45.0
Design & Engg	2.0	3.0	4.0	5.0	6.0	6.0	4.0	4.0	2.0	1.0	1.0	1.0	1.0	1.0	42.0
Services							6.7	13.4	6.7	13.4	8.7	10.1	8.0		67.0
Risk %														22.0	22.0
Warranty														45.0	45.0
Others	2.9	2.9	2.9	2.9	2.9	2.9	2.9	2.9	2.9	2.9	2.9	2.9	2.9	2.9	40.0
Total	6.9	7.9	8.9	39.3	67.8	122.7	127.4	68.2	97.9	104.6	45.0	27.9	13.9	71.9	810.0

Inflow

Details	M1	M2	M3	M4	M5	M6	M7	M8	M9	M10	M11	M12	M13	M14	Total
Adv. 5% (Contract)	50.0														50
Adv 5% (Drawing)				50.0											50
Supply - Progressive					41.2	82.4	164.7	164.7	65.9	123.5	123.5	41.2	93.0		900
Services													120.0	120
Total	50	-	-	-	50	41	82	165	165	66	124	124	41	213	1,120
Cash Flow	43.1	35.3	26.4	-12.9	-30.6	-112.1	-157.1	-60.6	6.2	-32.5	46.0	141.6	168.9	310.0	

As a result, the variance between the two different versions of cash-flow is reflected.

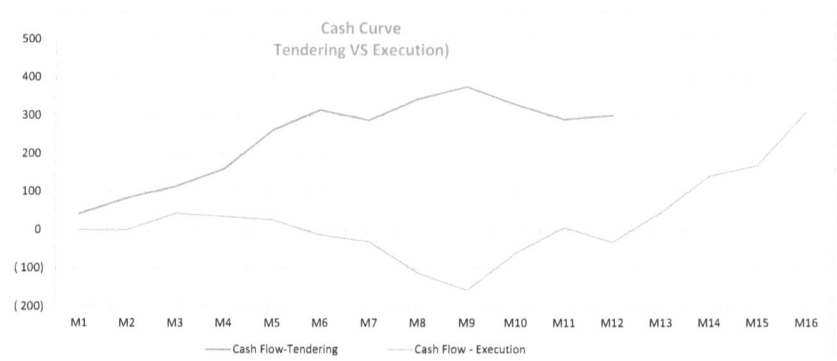

During the tendering phase, positive cash-flow was estimated throughout whereas actual cash-flow turned into negative during the execution phase for 6 months out of 14 months. Some of the reasons that caused negative cash-flow include:-

- Advance on drawing approval received in the fifth month of project commencement whereas during tendering advance it was estimated in the second month.

- Payments from customer were mostly delayed beyond the agreed credit period i.e. 15 days whereas XYZ Company had to honour its payment commitment to its vendors as per the agreed schedule, as no back-to-back payment terms were agreed. If the contractor would have made payment commitment back-to-back (i.e. releasing of payment to vendors only after receipt of corresponding payment from customer) cash-flow would not have been affected so badly.

- XYZ Company did not assess the customer's financial situation and past payment track records. All contractors or suppliers who worked with the Customer knew the fact that the company does not pay on time, as per order conditions. This market intelligence was missed on the customer payment pattern.

As a result, contractor-estimated cash-flow was incorrect and considering positive cash-flow did not envisage any finance cost although due to negative cash-flow finance cost 6.43 MINR was incurred.

Thus, a cash-flow must be duly assessed by a finance expert and due risk of delay or default payment must be considered in the risk register.

7. Double Loading in Costs

In case multiple business units or departments of an organisation are involved, then it must be ensured that multiple departments are not considering the same cost or risk provisions. Many clever managers may hide or keep risk provisions under the sleeves, which not only makes the proposal non-competitive but also an incorrect projection of the project financials. Thus, costs and/or risks envisaged by one unit or function should not be repeated or double-loaded. The same rule applies to all vendors or sub-contractors. For example, if you have awarded fixed-price-contract to the equipment supplier and it has considered the risk of raw material inflation, the same risk should not be considered by the other party. It can be managed through proper contract negotiations with vendors and double risk consideration can be avoided.

Contractor and its sub-contractors should work closely and clearly understand the scope, costs and risk distribution so that double cost and incorrect risk estimation can be avoided. Similarly, the insurance cost of supplies and services should be considered by either of the parties and not both. Such duplicate cost burdens should be avoided to increase competitiveness and agility.

8. Project Profitability

Executing projects may be the usual course of business for many organisations. Yet, the prime objective of *undertaking projects is to*

earn profit. Profit/loss means the differentials of earnings/revenue and expenses. In case earnings are greater than expenses, it will be called profit. Else, if expenses are greater than earnings, it will be called loss.

During the bidding or concept stage of the project, the contractor estimates the cost of execution and adds the profit margin to derive the project value. Unless there is a monopoly of the contractor, the profit margin is generally resolved on the basis of market trend or at par with the competition with minimum differential. These days, large projects' prices are decided through reverse auction or through T1L1 (technically superior and financially lowest) evaluation basis.

Due to volatile market scenarios, many contractors do refrain from long-term cost commitments. This is more common in infrastructure projects and contractors tend to prefer cost-plus basis model contracts, where actual costs incurred plus agreed margins are paid to the contractors. Thus, the accuracy of the cost is most critical. Failure to ascertain correct costs can lead to devastating outcomes.

Detailed analysis of probable risks and opportunities should be done meticulously. Depending upon the nature and complexity of the risks, it can either be passed on to customers, vendors and banks or absorbed if it can be managed through internal means. Cost escalation risks can be transferred either to customers with the provision of price variation clause based on market-linked indices or to vendors through firm price contracts.

A clear understanding of contract conditions and cost provisions are essential to safeguard project profitability. Onerous, unclear or unfair contract conditions related to scope, deliverables, Incoterms, penalties, liabilities, completion criteria, handing-over procedures, vague performance parameters, warranties, etc. can blowout the project profitability to negative.

9. Changes and Claims

Customer may delay or default in discharging their timely obligations in multiple ways, including delay in providing site fronts, approving drawings, making payments, changing scopes and specifications, assigning extra works, etc.; all such reasons may require additional time and resources, due to reasons not attributable to contractors. As a result, contractors incur additional unplanned expenses or prolonged engagements. This becomes tumorous if such changes occur at later stages of the project. Identification of such events and triggering variation orders or extra claims from the customer may help the contractor to defend the bottom-line and be instrumental to enhance project profitability. However, if the contractor is unable to define proper scope boundaries, it will miss the opportunity to register additional claims from customers; then it will add to their costs and expenses. Another important aspect to remember is that sometimes, additional scope requirements also crop up at the end, which holds back the total project completion or hand-over for months together. This not only drags the realisation of balance cash (retention payments) but also extends legal liability for a long period.

During execution continuous regular monitoring, review and update of project costing is the key to safeguard project profitability. It will help to highlight the probable risks and opportunities and take appropriate action before it is too late. A strong performance review system should be implemented to monitor the earned value analysis and cost to complete the project, as many costs incurred which are not recorded or provisioned, may lead to expensive surprises later.

10. Billing Methods

Billing methods play a significant role in deciding the contract prices, managing cash-flow and profitability of the project. Different billing methods are prevalent in the construction sector: –

Milestone – This is used in fixed or lump-sum price contracts where the number of milestones is agreed upfront between the parties for a defined scope. Upon achievement of the defined objective, bills are raised by contractors. Customer may add incentives for early completion and penalties for delays. These contracts transfer all risks and responsibility to the contractors.

Unit Pricing – This is common, where every item, deliverable or activity is billed accordingly to pre-determined unit rates of items. In such a scenario, a detailed Bill of Materials is fixed, with specified material, service, quantity, and unit rate. Risk of price escalation rests with the contractor. This causes conflicts while quantities are increased or reduced substantially or projects are delayed beyond agreed timelines, as price validity may not remain any more conducive. Many contracts may have provisions that in case the quantity variation (either upward or downward) exceeds the specified limit (%) of original scope, the rates of such material are refixed with mutual agreement between the parties.

Cost-Plus – This is a variable pricing method, where bills are raised considering actual costs incurred by the contractor for material, services, or activities, and thereby adding an agreed percentage or fixed fees of the mark-up of profit premium and overheads. This helps to ascertain the exact unit cost of each material, activities, or services. Risk of price escalation rests with the customer.

Time & Material – Mostly used where the complete scope is not clear or fixed. These are billed on an hourly or daily basis type of services. More prevalent in R&D, innovation, or software development-type of projects. Such services can be availed for certain short temporary services required at project sites. This is, however, preferred for small projects and as such not common in construction projects.

Firm fixed-price or milestone methods are most popular in construction projects, where billing and payments are made commensurate or proportional to the work completed. However, regular conflicts arise where quantities exceed the original Bill of Quantities agreed between the parties. Regardless of the reasons for changes in quantities, generally customers do not have adequate prior-approved budget to accommodate increased values, which affects performance and payments and also causes a delay in settlement of change order amendments and overall project closures.

One common mistake done is agreeing on the billing methods but not clarifying the payment methods, as the two are different.

Payment methods could include the following:

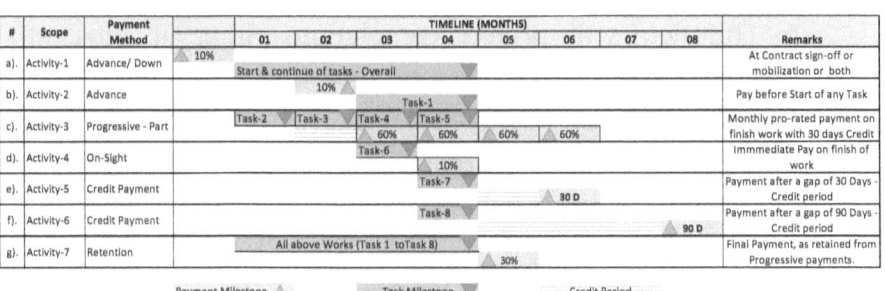

a. **Advance or Down payments** – are made before the start of activities. For example, 10% advance payments are paid to the contractor upon signing of the contracts or mobilisation of resources.

b. **Progressive payments** – are pro-rata or linear distribution of due amount, paid on a partial basis upon achieving the agreed milestone objective. Balance is retained by the customer and settled on completion or other agreed stages of the project. Most of the service bills are raised on running-account-bill-basis. It means each bill shows the current work, cumulative work carried and balance work pending till date. Many customers opt for progressive payments for supplies in running-account bill methods also.

c. **On-sight payments** – are made immediately by the customer upon completion of activity and receipt of invoices, supported by other agreed documents.

d. **Credit period** – Construction sector follows a practice of making payments within an agreed credit period, standard is between 15 and 180 days. In such cases, the contractor assumes the interest cost of delayed payment realisation. Good practices may involve bill discounting, where contractors receive the payments by adjusting the discounts or interest rates, for the period prior to credit period.

e. **Retention payment** – Part of the payments are withheld by customers from progressive claims and are released to contractors at project completion, either at substantial completion or final completion or any other intermediate milestone agreed between parties. In exceptional cases, customers may hold the retention payments until completion of warranty or defect liability period.

If timely payments are not made to the contractors for undisputed claims, it can hamper the pace of performance of the project and the contractor may seek recourse in the contract provisions to charge interests, suspend or terminate the contracts.

Finally, besides billing and payment collections, regular financial reconciliations with all suppliers, sub-contractors and partners are strongly recommended, so that in case of any mismatch, discrepancy or differences, the same can be resolved promptly. Each project carries out financial reconciliations by adjusting the due amount pertaining to liquidated damages, recoveries, variations, extra claims, insurance, disputed works, material shortages, etc.

All projects require adherence to the compliances and making due statutory payments and maintaining documentations thereof by all organisations related to taxes, labour welfare, fees and other charges. Any failure to fulfil compliance requirements is subject to penalties, interests and charges. These settlements must be mutually agreed between the parties before final completion or closure of the projects.

11. Taxes, Duties & Statutory Levies

Taxation of a construction project is a complex subject. It can pose various internal and external risks. All organisations need to strictly adhere to sets of rules, guidelines, codes, regulations, requirements during the construction phase to avoid violations or non-compliances.

Incorrect tax planning or non-compliance of taxation can trigger stern penal action against the defaulting party and may have a severe impact on project financials and legal liabilities. Correct taxation planning from the selling stage helps in timely compliance of taxation obligations. However, there is always an unknown risk of change of tax law, which can impact the project financials, if not managed well in the contract. There is no way to completely avoid risks as there are bound to be unknown factors that may arise at any time over the project duration.

Although the taxation terms and treatments are commonly defined in tender or contract as per applicable law. Usually, customers advise to consider the taxation in the project costing and expressly declare or disclose the taxes considered, so that in case of a change in tax law, suitable adjustments can be made according to the contract provisions.

Parties must agree upon as to which taxes, duties, statutory levies, exemptions, rebates, refunds, impeachments, etc. are assumed in the contract, with their clear-cut-off date of applicability, as any change thereafter may require proper treatment and adjustment. This is a common issue that if the projects are delayed beyond the original schedule and there occurs a change in tax rates or slabs, the parties may have a conflict. So, such scenarios should be duly considered and resorted to during contract negotiations.

Further, in case of any additional tax levy imposed during the contract period, the contractor should negotiate in such a manner that all such variation to taxes and duties should be to the customer's account. Terms of taxation, taxes applicable, required documentation for taxation, etc. should be mentioned in the contract with clarity to avoid ambiguity, conflict or violation.

Concludingly, managing costing, cash-flow, profitability, and claims are crucial to managing financial risks in projects. Uncertainties and changes can crop up at any stage of the project, from inception until close-out. Regular monitoring of project financials and taking immediate corrective actions help to minimise the damage and enhance the visibility and predictability of the project financials.

Chapter 5

SCOPE & TECHNICAL RISKS

"Well-begun is half-done"

– Aristotle

Factually, it is incorrect in the literal meaning that a good start implies the work is half completed, but a poor beginning may lead to frustrations, distrust, conflicts, or failures. This is true for projects too. Well-planned and well-started jobs boost the motivation of the stakeholders and chances of success.

Usually, there is a gap in the understanding of the 'scope' and 'requirements.' Rather, at times, these are used interchangeably and may lead to conflicts and confusions.

Project Scope is the description of all the work and activities to be performed to create deliverables in order to achieve the project goal or objective.

Project Requirement is the conditions, capabilities, or guidelines to be met or satisfied as per stakeholder's needs and expectations, fulfil contracts and specifications' requirement. The requirement includes scope, schedule, budgets, boundaries, constraints, characteristics, and performances to be satisfied.

Project scope can be described through a scope statement or by using breakdown structure (WBS). It helps to clarify 'inclusions' and 'exclusions' from scope. Scope also establishes the extent of boundaries or battery limits of the project. Sometimes, it may be easy to define the project scope but hard to agree and negotiate between the parties.

While defining the scope or participating in tender opportunity, all required inputs, information, data, and situations are not concrete and clear. Thus, scope is invariably defined with certain assumptions. These assumptions and estimations form the benchmark to prepare the baseline budget, schedule, and risks of the project. As the project progresses, the inputs, data, and situations become clearer or authentic, which leads to changes in scope. For example, site soil assumptions made at tendering stage can change when actual topology, geotechnical, seismic or other soil studies are made. Such inputs then force changes in original baseline inputs and consequently impact design and engineering.

Further, inputs received through tender documents or initially shared by the customer, related to layout, profiles, soil, approach, transmission system, etc., may undergo changes during contract-formation or execution. Therefore, while or before negotiating the project scope, the team must understand and analyse its constraints and dependencies, which may impact the performance and/or outcomes. Changes in scope or technical parameters may have an impact on project timelines, budget, resources, and risks.

Some project scope-related risks, which can lead projects to failure are:-

    a. Lack of understanding of project scope, requirement, and expectations,

    b. Incomplete or generic scope definitions, specifications, or tender terms,

    c. Under-design or over-design of system equipment or plants,

d. Casual or no approach for risk identification and mitigations,

e. Unaware of risks emanating from the use of new technology or equipment due to untested, untried methods or missing past lessons,

f. Failures to correctly assess the impact of additional requirements,

g. Too many or frequent scope changes or scope creeps,

h. Stringent quality, performance and reliability requirements,

i. Inconsistent or conflicting specifications or contract documents,

j. Poorly defined completion or acceptance criteria,

k. Scope change control mechanism – either not defined or not followed,

l. Overcommitments, etc.

If the above points are understood, analysed, and considered thoroughly, then the probability of project failures can reduce drastically. Some of these risks are elaborated below with commensurate mitigation levers:-

1. Project Scope Risks

Scope risks are related to specifications, performance, guarantees, reliability, functionalities, protections, technology, etc. used for material, equipment, and plants. So, first time right, the scope should be defined carefully with precision and accuracy and associated risks and dependencies understood, within the constraint of time, cost, quality, and resources. Customer should also define the exact deliverables, layouts, boundaries, specifications, battery limits of product, services or results that are going to form the basis of the project scope. Once the scope is defined, documented, understood, and agreed between the parties, thereafter it should be put up for approved by decision-makers and/or contracts are signed by parties. At times, the project scope is defined with hidden requirements or implied expectations by some customers, which cause conflicts.

Project scope has to be executable within constraints. Although some projects' scope may look normal or business as usual, but considering the constraints (timelines, budgets, quality, resources and risks), this may not be feasible to accept or execute.

For example, a customer may require the construction of 220kV gas-insulated switchyard (GIS) to be completed within six months. The scope may include acquisition of lands, import of GIS switchgear or other key equipment from other countries. So, under the time constraints, these risks may not be mitigated. At times, the import of GIS switchgear from another country may take over six months.

1. **Scope reduction** – Many customers may negotiate larger scopes, quantities, or volumes initially during tendering or contracting stage, as part of their negotiation strategy. This helps them to attract higher discounts or competitive rates. Such customers may integrate the requirements of multiple projects (ongoing or futuristic). While concluding the orders or taking actual deliveries, they reduce the scope. Upon scope reduction, this affects the financial and operational feasibility of apportionment of the fixed costs and overheads with reduction or limited scope. *So, in such cases, the contracting organisation can use levelled or slab-based pricing so that the risk of reduction is mitigated.*

2. **Risk of long-term scopes** – Many customers may demand rate contracts or framework arrangements and bind contractors to serve the agreed scope for a long time. In such cases, the costs, overheads, and profits are apportioned based on a larger scope and longer durations. Besides the risk of reduction, another major risk associated with such contracts

is of price parity and validity. If firm prices are agreed for a long period, the risk of uncontrollable exposures of inflations and fluctuations in commodity, currency, interests and labour may not be assumed correctly. *So, in such cases, the contracting organisation should negotiate the price variation clauses by fixing a base date and indices. Any change thereafter shall be adjusted suitably at the time of transaction. Indeed, many a time, the project involves commodities which are highly volatile and suppliers or contractors can agree to the conditional prices, with an opportunity for price revision at the time of actual transactions or shipments or payments.*

3. **Slips in forecasted numbers** – Most projects start triggering changes from contract-formation itself and continue throughout the project tenure. Thus, project scope and changes must be closely documented, controlled, and managed discretely. It is revealed that multiple scope changes severely impact the project outcomes and lead to a significant departure in its projected numbers, i.e. "As-Sold" versus "As-Delivered" state of the projects. *So, projects should be monitored based on earned value principles preferably with time and cost. It means that scope discussions should not be done in isolation, without evaluating its financial implications. Any abnormality or gap can be quickly addressed or escalated for resolution by the project team.*

Deliverables of small projects may be agreed or defined informally, but for large and complex projects, more rigorous and robust risk management processes and experts must be involved from the beginning of the project. This supports decision-makers to make informed decisions based on risks involved and risk appetite of the organisation. *Strange but true, sometimes risk identification or risks analysis feels like frustrating or onion-peeling while experts get into the nuts and bolts of the scope.* This not only helps to define scope deliverables clearly but also to discover the correct risks and make appropriate mitigation.

4. **Optional scope risks** – Many customers may include the additional optional scope, items, or activities in the contract. These optional scopes may not be required immediately and not included in the total contract price. However, upon demand from customers, contractors are bound to execute the same on the rates or prices provided to the customer. The customer may invoke the need for such optional activities at a later stage. If such optional prices are provided without due diligence or back-up commitments, it may be unviable or become a point of conflict between the parties, if the need arises to execute. *So optional scopes or prices must be reviewed carefully or refrain from including in the main contracts.*

5. **Expert resource risks** – Scope risks also creep in due to non-involvement of specialist or key contributors. In such cases, scope problems become visible late when core or key members join the team. These experts or key contributors (internal or external), team or agencies, may include technical experts, end-users, customers, sub-contractors, suppliers or other stakeholders (present or potential). *So, proper stakeholder planning is to be made to ensure involvement and availability of relevant expert at the right time to add value, early detection of risk and mitigation plans.*

Nonetheless, involving experts may not zero down risks or grossly eliminate errors, but it will certainly minimise the probability or impact to large extent. As experts also have to be dependent upon certain assumptions and situations, which are subject to change.

At times, in order to please customers or conclude the deal faster, some professionals may make higher commitments than assumed in the costing. Such gold-plating inflates the scope with financial burden. Any such overcommitment must be documented in the risk register.

Furthermore, those who are negotiating scopes, solutions or contracts with customers should also have adequate domain knowledge, relevant experiences in industry, products, projects and solution. This helps to negotiate a robust solution and prevent unreasonable exposures and liabilities, once contracts are executed.

Following inputs give a clear direction and benefit to define the scope and reduce exposure:

a. Alignment with the organisation's Vision, Mission and Strategy

b. Clear understanding of customer scopes, needs, deliverables and expectations

c. Project site locations and local conditions are surveyed and assessed

d. Understand Compliances (Environmental, Social, Governance)

e. Reasonable assessment of attractiveness and feasibility of opportunity

f. Clarity of applicable standards, codes, and current industry practices.

g. Competition mapping – What do competitors and market have to offer?

h. Customer decision criteria (Hierarchy, preferences and evaluation mechanism)

i. Trends and patterns of past customer projects

Above aspects are not complete but are shared to trigger ideas so that project teams can explore or analyse in view of the relevant exposure of their project. It allows the decision-makers to make intelligent choices and decide if they had the risk appetite to absorb and ready to accept the level of risk identified. In case the risks are higher than the expected rewards, decision-makers can drop the project and invest their time and energies on other opportunities.

It is easier said than done, in the initial stages when too little is clear, many leaders may have two options – (a) accept the risks associated with partial inputs or assumptions, or (b) abandon the project. However, a team can explore the third option. Rather than blindly accepting the unclear scope due to lack of data, the *project team* defines the specific requirements. These requirements shall be based on previous projects, assumptions, experiences, and current understanding of the problem and accordingly negotiate with the customer *conditionally*. Any changes arising with the passage of time shall be mutually contemplated and necessary changes or revisions are incorporated, with commensurate time and cost compensation.

Therefore, '*correct understanding of scope*' and '*correct identification of scope risks*' boost the probability of success and feasibility to take correct decision accordingly.

2. Assumption Risks (Ambiguity Risks)

Assumptions are the factors, situations, speculations or events that are considered to be true, real, or certain. Assumptions pose potential threats and risks. In the early stages of the projects, technical teams anticipate or simulate many situations and data that are assumed to be true and accordingly make decisions. These assumptions may be based or backed up by previous projects, experiences, references, inputs received or other industry norms. Assumptions can be related to design, scope, dimension, sizes, layouts, costs, productivity, technology, market intelligence and so on. The risk with assumptions is that there is a high probability that these assumptions undergo changes while actual information starts flowing.

Project scope largely relies upon assumptions, estimations, and simulations. Assumptions made at the tendering stage have a high likelihood of changes at the contract-formation stage and thereafter during execution. So, thorough and rigorous review of these assumptions and estimations are crucial for projects at every stage.

The assumptions are as correct as the source and authenticity of data itself. If the sources or data itself is inaccurate and questionable, it will lead to mere guesswork, which is very risky for projects. This could trigger a lot of changes, risk, and consequences. Though change may be an opportunity for the contractors, but if the project agreed is fixed-price contract, such risks may have catastrophic impacts.

For example, if the tender specifies the site soil as alluvial or red, one can plan for drilling or underground cabling activities. But, after actual site survey, if the soil is found to be changed to mountain or hard rock, any assumption for drilling or underground cabling shall go for a toss or even impossible to work. So, appropriate methods of drilling or cable laying are to be considered.

Similarly, change in site location from normal plain land to any coastal, desert, or hilly terrain can pose a different set of risks and challenges and require appropriate selection of construction methodology. So, any advance assumption may undergo significant changes based on the availability of actual or correct details. Furthermore, if prices offered or agreed are firm, regardless of changes in soil or site, the consequences of cost and efforts can be devastating.

All assumptions should be documented and made part of the contracts. Another classic example of assumption is that if an equipment is specified to operate on a particular temperature level or voltage level and subject to periodic preventive maintenance, the same should expressly be specified in the contracts or O&M Manuals provided to the customer. However, if the customer fails to maintain the required temperature or voltage levels during operations, then the equipment may underperform, trip, fail or even cause damage to other components of the system. The biggest risk is involved if the contractor fails to provide the O&M Manual or does not specify the exact battery limits or prevention cadence; then it is bound to void warranty obligations and may also cause big disputes in case of defect or failure of equipment or plant.

For example, suppose the land parcel for the project is not finalised by the organisation, but the architect or designer is required to prepare plant or equipment layouts in the project. While the actual land or layout is unknown, it is difficult to assume or ascertain its coordinates, sizes, shapes, profiles, and soil inputs. So, any assumption by technical experts may undergo change which can significantly causes rework and impact on time and cost of the project.

Similarly, if power evacuation voltage levels of grids and transmission systems are unknown and designers prepare the technical specifications for electrical equipment, then it shall be based on assumptions with a high risk of changes. Thus, the information is the key to correct decision-making by the concerned stakeholders. In such cases, neither customers may approve drawings nor can contractor place orders or accord manufacturing clearances.

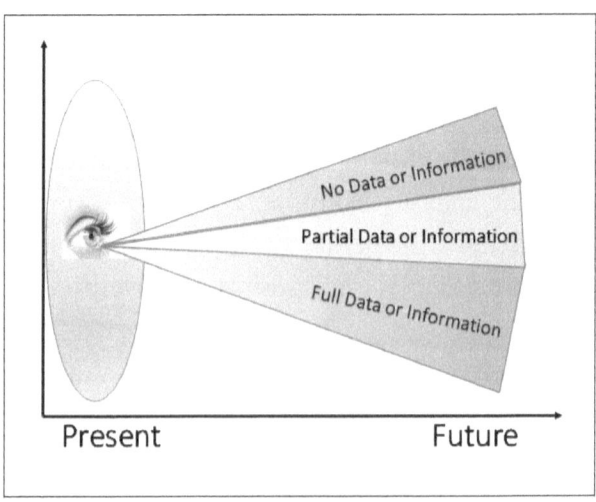

Besides the above, there are many similar assumptions made by project team like:-

- Project shall complete as per our schedule, assumptions and resource plans
- All approved suppliers shall deliver the material as per our time and budget estimates
- Construction sites shall get adequate required (skilled and unskilled) resources as per deployment or productivity plans
- Training imparted shall completely eliminate workmanship defects and issues
- All construction material for civil works, machines and tools shall be easily available within the local vicinity of the project site
- Manpower cost of the project will not increase during project execution
- Site setup and infrastructure costs shall be fixed and flat and shall not increase
- All parties have clearly understood the scope and requirement and there are no changes envisaged

In case land acquisition is included as part of the contractor's scope, the set of risks and challenges shall be very different, as compared to the projects where land is provided by the customer, as land attracts a lot of risks related to the boundary marking, acquisition, conversion, right-of-way, compensation, development, and litigations, etc.

Imprecise estimations can lead to wrong design, engineering, or Bill of Quantities. These will ultimately cause unexpected changes, claims, or conflicts later.

Furthermore, as an integral part of the project scope, most construction projects need approvals and permissions from various government agencies like ministries, forest, defence, highway authorities, land, revenue authorities, DISCOMS, utilities, CEIG, CEA, pollution control boards, labour, corporations etc. These take a lot of time, coordination, and charges, which are usually not budgeted correctly in costing. If timely approvals or permissions are not secured, this can become a deterrent to the project construction work or completion.

3. Scope Changes

Scope changes are normal, unavoidable, and universal. Perhaps, changes cannot be eliminated from projects but can be managed by applying appropriate principles of change management or risk management. Despite having well-defined scope, understanding, discussions, and negotiations between parties, projects may still encounter changes in their lifecycle. It is a myth or disbelief that a project shall not have any changes in its lifetime.

The scope changes may emanate at any stage of the project and from any situation, party, stakeholder or statutory reasons. Changes mean any addition, deletion, reduction, replacement, upgradation, or modification, etc. in the originally defined scope (baseline) of work. As a result, scope changes can have a light-to-significant impact on deliverables or commercial aspects of the projects.

Changes may affect any aspect of the projects, including design, technologies, material, makes, features, tests and inspections, methodology or so on. Changes can be as simple as a minor change in the shade of the paint of a building to as complex as a complete overhaul of new technology or design, which impacts heavily on budget and schedule.

It is not recommended to reject or prevent scope changes because the project environments are uncertain, dynamic and keep changing, which can force the projects to undergo changes. Changes may cause extra cost burden at the time of occurrence but may be beneficial for a long-term perspective like switching or upgrading to the latest technology, etc. Indeed, many situations or conditions can trigger a change in the project like site conditions, evolving technology, competitive market conditions, need or entry of new stakeholders, statutory laws and regulations, and other socio-political aspects.

For example, in a large power plant, if the metering clause is not defined in the tender, but at the time of drawing or metering methodology approval, the customer specifies exact 0.2 accuracy class, this can be a change from the original specification, which will have appropriate impacts.

Similarly, if the metering or protection system is assumed based on specific standard or State Utility specification, but later if the changes are made in the standard or as per different standards or Central Utility compliances, the same may be qualified as a request for change.

As one cannot restrict or reject the changes in scope, the only viable solution is to have a proper change control mechanism in place. Absence of a well-defined change control mechanism may be the key hurdle or risk to the project. If the parties fail to establish a change control procedure, then it will continue to fulfil the whims and fancies of never-ending customer expectations, which will be practically difficult to honour throughout.

Uncontrolled scope changes can turn out to be a killer for the projects. However, the objective should not be to stop the scope changes or scope creeps but to establish good governance to manage the change process. Any change request triggered by a contractor or supplier or consultants must be substantiated with adequate evidence, justification, and cost/benefit analysis. Nonetheless, the need for scope changes can be triggered by customers, contractors, changes in the law, technology or any other reason.

The scope change mechanism should also be balanced between flexibility and control. In case the mechanism is too cumbersome, the stakeholders shall ignore it or be reluctant to follow. As a result, the project will not receive valuable inputs from its stakeholders. On the contrary, if the mechanism is too easy, it will face too many changes which will not follow adequate thoughts and due evaluation of its merits and consequences. Thus, a proper process with defined authorities, timelines and accountability should be built into the process to meet the pace of changes with proper due diligence and evaluation.

Most changes are discovered either at the early stage or completion stage of the projects. Notably, frequent, and substantial changes in scope can add complexities to the project and demotivate the project team. This even makes the relationships sour and conflicting among the stakeholders.

4. Documentation Risks

Inherently, all contracts, projects and businesses have documentation risks. All businesses, projects, trades, agreements, loans, land and infrastructure, considerations, etc. are based on documents only. It is sheer ignorance if someone believes that there is no documentation risk in a project. It can be an unknown-known risk. In case of any issue, disagreement, change or conflict, all parties look into the documents and their positions. Documentation risk means that the documents are legally valid and enforceable under a particular contract or recognised under agreed law, if required.

There can be multiple risks associated with documents. Some documents are required to be legally enforceable and some for generic utility, reference or for discussions. However, to avoid documentation risk, it should ensure that: –

  a. Documents, contracts, or communication is legally valid, binding, and enforceable.

b. Documents correctly reflect what was discussed and agreed between stakeholders

c. Documents to cover complete scope, scenarios, situations and assumptions

d. Any cross-reference to related documents, rules, standards, laws, etc. are expressly defined, understood and order of precedence clarified

e. Documents are executed by authorised persons of the parties.

After discussions, understandings and analyses, the project scope can be mutually agreed between the parties. If the same is not documented properly, it may be detrimental for the project, as with the passage of time, oral understanding, interpretation, people, scopes, situations may undergo multiple changes. Scope documents should expressly describe "Must be" (Essentials) and "Would be" (Wants) requirements.

Sometimes, initial tender documents are released with generic specifications, which do not conform to any specific standard or does not specify exact standard compliances. Such specifications become open-ended and difficult to comply with or conform to any specific standard and can create conflict. Thus, while finalising the contract, the parties should agree on the conformance to specific standards like IEC/ISI/IEEE and do not leave to be open-ended choices. This becomes more difficult and complex in international biddings because the scope of standards is bigger and complex.

Further, the team should clearly agree in writing on the following key aspects:

- Definition of every activity, package, key deliverables
- Inclusions and Exclusions of Scope
- Completion and acceptance criteria of every task, activity and deliverable

- Supporting evidence required to substantiate its deliverables
- Inter-dependencies with timelines on each stakeholder
- Basis of assumptions or estimations and source of data considered
- All referenced layouts, drawings, specifications or makes agreed
- Features, functionalities, reliability, performance parameters and criteria, etc.
- Formats, checklists and frequency may be used for reviews, inspections, audits, etc.

To the extent possible, teams should agree at the beginning about project acceptance, completion, handing-over-taking-over (HOTO) deliverables, process, procedures, checklists, and documents requirement. This shall smoothen the project execution, completion and exit from the project.

For example, a company sub-contracted a job to a local fabricator for the supply of galvanised steel structure for module mounting for a solar plant. However, they missed to document about the chemical and mechanical properties of the material required. Considering the high volume of steel structure required, the customer could not inspect the same and waived the inspection. As a result, when the customer carried out a random inspection, it failed in intermediate lots. This becomes a big roadblock for both the organisations, how to ensure the conformance to the contract specifications and reliability for the lifetime of the project, as both the scenarios have their merits and consequences.

Similarly, a turnkey contract was awarded to a contractor for the construction of the control room in a substation. However, the contract documents and drawings approved were silent on the supply and fixing of the sanitary and electrical fittings in the control room. This was assumed by the customer that this is a common practice in the industry and implied that the contractor would furnish the same. However, during execution, the contractor did not supply as they had not provisioned in their costing; as a result, this became a point of conflict and then led to scope change. So the risks associated with all fittings, accessories, consumables, spares, etc. must be properly agreed and documented and not left to any assumption.

5. Scope Changes Versus Defects

One of Murphy's Law says, "*Whatever can go wrong, will go wrong.*" Quite often, any deployment of new technology or solution does not function as smoothly or straightforward as it is predicted or projected during the selling phase. Even, a regular and reliable product or solution may encounter faults or fail to operate as required. There can be various reasons, but ultimately the project is impacted and may require rectifications with more time, cost, efforts, or resources.

On the prima facia, project scope risks are related to its deliverables. Among the total risks encountered in projects, the project scope and costing risks account for nearly half of the total impact on the project. So, managing scope and costing needs the highest priority in any project.

Broadly, project scope risks may be categorised as **Scope Change Risk** or **Scope Defect Risks**. Scope risks pose significant exposure in projects. Many professionals mix and describe the defects as changes. The scope risks arising out of defects are generally considered as *unknown risks* and lead to rework or replacements, leads to additional financial burden to the contractors and may be called cost of poor quality (COPQ). With thorough attention on scope definitions and robust quality processes, such scope risks can be minimised substantially. Unless intrigued by unknown risks emanating from the deployment of new technology or experiments, most of such defects can be avoided and money earned is saved.

a. **Scope Change Risks** fall in any of the following groups:-

- *Scope creep*: Enhancement or additional scope of work (new activity, item, equipment or services, etc.) instigated during execution, which can have an impact on time and/or cost.

- *Scope Assumptions*: Changes in data, inputs, needs or expectations from what was '*assumed at project start-up*' versus '*actual in execution phase.*'

- *Differences*: Changes (addition or deletion) in agreed scope, specifications, or activities during project execution.

Albeit, all the above incidents cannot be perfectly anticipated at the early stage and cause an impact on time and cost of the project upon occurrence.

Scope changes may play a dual role, as some changes can bring in opportunities for contractors to make additional revenue, save costs or enhance profit margins, whereas some changes can be deterrent and expensive as these form an integral part of turnkey or fixed-price contracts and contractors are not entitled to additional benefits. Some organisations emphasise on continuously identifying the scope creep to improve their contract value, profitability and extended engagement with customers.

b. **Defect risks** fall in any of the following groups:-

Construction projects involve intricate multiple activities and scope of works. These have a high dependency on many uncertain and dynamic factors. Thus, it has high exposure and potential for defects. Defects are caused due to inappropriate procedures, excessive variability, damages or excess variable material, worn machine parts or human mistakes.

Defects in construction projects are basically the errors, deficiencies or insufficiencies which may come out due to poor controls or weak processes related to design, drawings, material, workmanship, manufacturing, or environment. This may lead to failure or deficient performance of the equipment, plants, or structures from expected levels. The worst defects can cause a threat or damage to persons or properties.

Defects may refrain the material, equipment, structure, or machines from use, perform, or function as required or expected. Equipment and machines also pose risks of damages or defects due to jolts during transit or handling. Even automation systems like automation or SCADA system may not be configured well or fail to produce desired results or reports. These can have a lack of features, functionalities, redundancies, reliability, or customisation issues, which can be defects too. Likewise, civil constructions can have enormous defects related to inferior material quality, workmanship, use of wrong tools and practices, aesthetics, seepages, leakages, or many others.

Defects are classified as patent or latent. Patent defects are generally visible or obvious and can be easily detected during inspections or supervisions. These are easily found by field teams without much effort. Latent defects are invisible, concealed or not readily detected during normal inspections or supervisions. Patent defects are easy to find, easy to fix, as accessing and repairing are not that invasive. Latent defects are hidden, difficult to detect and expensive to restore. Many latent defects may arise during or after defect liability period too. Such defects can be broadly reviewed as:-

    a. Design defects – professional failure to produce accurate outputs as per design, sketch or specifications and caused due to errors or omissions.

b. Material defects – arising due to the use of poor-quality raw material, use of damaged material, poor manufacturing processes or weak testing or inspection procedures. Sometimes, such defects are encountered at a later stage while operating the plant or after using the building for a long time. Rectification of such defects can be expensive and at times customers may have to live with it.

c. Workmanship defects – Such defects occur when construction work is not carried out as per correct and approved version of drawings, specifications, use of improper tools, documents or unskilled professionals are deployed instead of skilled workers. These defects may be caused due to changes, deviations, inadequate attention, process inadequacy, complexity or uncertainty. Such defects can be simple issues of aesthetic looks to structural integrity problems. Allocating liability and determining the root cause of such defects can be extremely cumbersome and expensive, so also the consequences. Examples are finishing works of plastering, wrong concrete mix, concrete finishing, soil compaction, waterproofing, painting, fabrication, cable laying and wire dressing, fixing of various appliances, coverage of gaps, etc.

Civil defects may also be related to design and engineering, structure, soil settlement, design mix, curing concrete slabs, building roofs, and finishing, etc. Mechanical defects may be related to structural work, plumbing, HVAC, etc. Electrical defects can be related to specifications, sizing, performances related to various equipment including cables, panels, transformers, modules, inverters, turbines, lighting & protection, etc.

To conclude, everyone in a project is responsible to minimise the construction defects, as the project is a collaborative teamwork. A lot of proactive measures can be taken to prevent or decrease the construction defects. The organisation should emphasise to implement good contracting practices, quality management system, quality processes (at manufacturing, processes, logistics and field construction). Construction defects can even turn a project upside down. Quality control programmes, communications, documentation, and continuous monitoring during construction are the easiest ways to prevent bigger damages and enable project team to meet the objectives.

Chapter 6

SUPPLY CHAIN RISKS

"Everything is negotiable. Whether or not the negotiation is easy is another thing."

– Carrie Fisher

Supply chain is ripe with risks. Due to disruptions in the supply chain, companies may lose hefty money, which may lead to cost volatility, non-compliances, damaging incidents and reputational perils. Supply chain disruptions could cause due to a variety of reasons like capacity issues, inventory problems, physical damages, social-political disturbances, natural disasters, social commotion, strikes, agitations, and other known or unknown risks.

Supply Chain Management (SCM) is a multi-tier system where a large number of suppliers, vendors, agencies and contractors are competing globally for all type of products, services, and logistics. There is fierce competition at every stage from raw material sources to producing the finished products, delivery services and even thereafter. The complexity and diversity of supply chain scope is intimidating. Practically, it is difficult to ascertain the exact and gross supply chain risks that are associated or can affect projects. Thus, thorough system, speed and spirit are required at each level to drive this value chain incessantly.

The outlook of a customer is not limited to cost and delivery, but also to ensure durability, reliability and ease of operation of material or plants, for its lifetime. All these are possible by discussions and negotiations skills. So, if an organisation is missing smart negotiation experts in SCM team, then this is a key risk for projects.

The supply chain is highly dependent upon expert negotiations. Organisations can benefit from an expert's negotiating skills by seeking simple discounts on safety gadgets (helmets, jackets, shoes), reducing delivery timelines or can get into the most complex negotiations of purchasing high-value capital equipment and plants or infrastructure projects. Negotiations are not only limited to discounts or lowest values of material purchased but the total cost of operations (TCO). This may include competitive prices, best deliveries, packing and delivery services, superior quality, add-on features, spares and replacements, and other aspects.

Supply Chain Management ("SCM") is actually a long lifecycle management of goods, services and logistics from raw material producers to integrators to customers or end-users. It manages the flow of material, information and cash among producers, suppliers and customers. It involves a large number of simple-to-complex processes, streamline activities and inter-dependencies to monetise maximum value and gain competitive advantage in the *marketplace.*

Risks related to the supply chain starts from the identification of suitable suppliers or partners, who fulfil the qualification requirement, are truly interested and serious to work with customers. This may be assessed through the interest expressed by suppliers or sub-contractors in collecting the tender or request for proposal (RFP) documents and submission of their offers. There are chances that many suppliers or sub-contractors may not find your tender or RFP documents conducive to their business models, experience, or risk appetite. In case many parties are not willing to participate in your tender, then it is a risk of not getting true market prices and invoking fair competition.

Every organisation looks for quality and reliable partners so that any of your partners does not back out or leave the transaction or project in between. Some of the risks associated with the supply chain include:-

1. Supply Chain Organisation Risks

One of the key pillars, that ascertain the success or failure of large and complex projects, is the maturity and stability of the supply chain organisation. The supply chain is the backbone of the projects. Therefore, selection of right sub-contractors, vendors, consultants, logistics agency and other partners is the most crucial risk for projects. This can be better addressed by putting in place a robust supply chain organization with adequate authority and responsibilities.

Defective selection process or defective hiring of partners can turn out to be expensive for a project or organisation. Failure or major default of key partner(s) can have an adverse effect on the project schedule, financials and may jeopardise the project objective.

While establishing the supply chain mechanism, it must be ensured that the selected partners have adequate systems, processes, capacities, capabilities, and expertise in discharging their responsibilities. It is of the utmost importance to assess the maturity of the partners in terms of its **people, processes, plans and performances.**

The consequences of selecting a wrong or incompetent supplier can be devastating. A wrong supplier or partner can derail the project from the track and may lead to frustrating consequences of abnormal delays, financial losses, conflicts and disputes, legal and regulatory actions even and goodwill damages.

Therefore, a robust supplier selection process and control mechanism is to be established from the beginning of the project. It is more than merely defining its quality audits which measure compliances with procedures and regulations. The true objective is to assess the capabilities of the partners and overall risks associated with their performance in fulfilling their commitments. As such, the performance of suppliers is not restricted to the progress of work, but also ensuring the right environmental, social and governance compliances (ESG) and ethical standards adopted by them.

So, while assessing the suppliers' or sub-contractors' creditworthiness or qualification, the customer should also gather the feedback or market intelligence about their own creditworthiness. In case, some suppliers or sub-contractors have bitter experience, then this message may be known to other suppliers and sub-contractors as well. As a result, most vendors may not provide competitive or serious offers, which becomes a challenge to identify the right partner.

Another aspect is that the terms proposed by your organisation, as a customer, should also be at par with prevailing standard market conditions. For example, if the standard market norm is to provide 10% of advance payment then your tender terms must be the same or similar to those only unless there is some special incentive or benefit available to partners. This is important because payment terms become a serious contributor to their cash-flow risk and hence this may drop their interest in participation, if not viable. If such conditions or issues are not resolved by SCM experts, your organisation shall run the risk of not an attractive adequate competition or correct pricing for your tenders to match your budget levels.

Besides the above, many more such points can be assessed by a matured supply chain team with proper feedbacks and field assessments. Some of the questions which every supply chain professional should ask about their potential vendors, sub-contractors and partners:

- Have they understood and can they cater to our scope, specifications, deliverables, requirements, timelines, quality, and other conditions?
- Do they have adequate capabilities, capacities and competencies to execute?
- Can they discharge their obligations, without failure in their commitments?
- Do we have back-up plans for critical vendors or suppliers?
- Can we optimise logistics of product, services and shipments?
- Do we have all processes in place and accountability clearly established?
- Have we considered reasonable financial risks and loading?
- What is our own market position or goodwill among competitors and partners?

It is noteworthy that leading suppliers can be engaged by your organisation and your main customers or end-users. So, it is important to note that the same set of suppliers may be negotiating with you as well as your customers directly, which can bring conflicts and undue competition.

The identification and selection of the partners become more challenging as supply chains operate from distinct markets and in diverse geographies. While selecting, a customer must carry out the adequate risk assessment and due diligence before clinching the final deal. This may include personal visits to supplier's factories, their projects (constructed or under execution) or their offices for a correct assessment of its cultures, processes, qualities, deliveries, and commitments. A new supplier or partner should not be decided merely based on its paper proposal, presentations, indoor meetings with their salesforce, who are good communicators and professional to sell anything. Many construction teams face major concerns because of over or unrealistic commitments made by frontline team during the selling phase. This not only creates confusions and conflicts but hampers project performance. A good practice is that the cross-functional experts of both sides (customer and supplier) be involved prior to contract-formation, for assessment of their respective scope of work and associated risks, though, final price negotiations may be held with a selective group of decision-makers.

2. Supplier Relationship Risks

It is all about relationships as to how well one manages or treats its partners—customers, suppliers, vendors, sub-contractors, and other service providers. If any organisation commences the relationship with its partners with pre-conceived notions, adverse mindset, or ego, it will only bring constant competition and conflicts in the relationships. Whereas if an organisation starts a relationship with the positive intent of partnerships with win-win objective, then the relationship is going to be long-lasting. Eventually, sometimes perceived win also plays an important role in negotiations. The customer-supplier relationship should never be based on the philosophies of finding faults, defects, letting down, assigning blames, etc. By implicating or blaming others, we do not solve problems, but end up spoiling relationships and endangering project objectives. Rather relationships should be reciprocal and complementary to each other. It should be based on a collaborative and collective approach towards resolving issues, challenges, or risks. It is far easier, better, and faster for the organisation to achieve objectives when relationships are congenial and harmonious. Such relationships not only support to achieve better performance and results but also make the work environment healthy and conducive for working teams as well. It does not mean that an organisation should not protect its respective interests or digress from the objective. But the same is still possible by maintaining ethical and mutually respectful behaviour. It is all about ethics, values and soft issues.

3. Makes & Origin Risks

Frequent anomalies and issues arise in a supply chain related to makes, models, manufacturers, and origins, from where the material can be sourced or supplied. Thus, parties must freeze the same during the contract-formation or initial kick-off meeting itself.

Customers should establish procedures to ensure that all purchases of material, equipment, machines, tools, spares, and consumables are made from pre-approved sources. Further, if a supplier has manufacturing units at multiple geographies (within India or outside), then the exact origin or place or unit must be agreed, as the type tests, inspections, standards, logistics and compliances may vary substantially.

As contractors are required to supply the material from pre-approved vendor lists (AVL) or supplier lists (ASL), it is important that such vendors, suppliers, OEMs, should be approved by the customer, involving their relevant cross-functional experts, so as to avoid any ambiguity later. Organisations may qualify suppliers after due assessment of their credentials or visiting their manufacturing works, project sites or offices. They may include experts from R&D, Product Management, Design, Quality, Operations, Supply Chain, Logistics and HSE, etc. At times, the supplier selection process may be time-consuming, multi-staged and complex. So, if multiple new vendors need to be assessed in a project, it can become an impediment or cause delays, so proactive actions or reasonable time and resource be planned at the early phase itself. Sometimes, even the process of vendor selections itself could be complex and need revamping to meet the objective of time and volumes.

Whether the material is purchased directly by the customer or indirectly (through a contractor) from any supplier (manufacturer), the buyer must put in place clear, unambiguous and robust procedures and methods for qualifying, dis-qualifying and re-qualifying of suppliers or sub-suppliers. Many organisations deploy dedicated teams for continuous development of new suppliers and vendors and performance assessment of existing suppliers.

The procedures set out should include the acceptable quality standards, methods and mechanism associated with the level of risk to the supply chain. Selection and frequent evaluation of suppliers and partners are crucial for the projects and the organisation. Based on the evaluation, the level of the controls can be established which can help to correctly identify and mitigate the risks associated with the supply chain.

Some contractors may not be very alert, matured, or vigilant about nuances of tender or contract requirement during award or negotiations. They presume that all makes, models, manufacturers of material shall approximately cost the same or may be managed within cost. However, while they start placing orders on their vendors, the suppliers (sub-suppliers) tend to opt for make or manufacturer which is the most economical. Unless expressly agreed in the contracts, the customer may also seek specific make of material from the agreed list, which can turn out to be relatively costlier. Many small suppliers or manufacturers may offer shorter delivery but compromise on quality, performance, features or TCO, which are not accepted by the customer during execution. Key risks associated with makes are linked with the costs and delivery timelines. Most proposal managers consider the lowest cost from an approved make list, to be competitive or lucrative. In case the customer demands a specific make (from approved list), the supplier bears the financial hit or creates conflict.

This is very common that initially contractors may accept the order or contract without evaluating the available material makes but later seek approval for new vendors for various reasons. There are mistakes made by contractors to delay the ordering to its vendors, which become unviable later. This forces the contractors to seek additional make approval, which is economical and justifying their plea based on crashed delivery or add-on features. Such last moment request for a new make approval puts the customer in a dilemma, although the customer has the discretion to reject or approve the new make. However, if the customer accepts the new make by protest or half heartedly. This may look beneficial to a contractor in the short term but certainly harms the relationship and trust with customers, as customers understand such tactics of the contractor. Therefore, makes, brands, models, manufacturers, or sourcing origins, etc. must be clarified or confirmed and agreed at the early stage of the order finalisation. Any new request should be invoked only if something is not agreed initially or an absolutely uncontrolled situation arises.

4. Material Inspection Risks

Interestingly, supply chain management, quality management and risk management are tightly interwoven in a project. Regardless of the organisation's maturity or advancement of processes, there will still be defects and deficiencies faced by the project team. Due to inherent uniqueness of projects, it is risky to precisely comply with all technical specifications, quality levels and standards. Recurrent defects or failures in material or workmanship may lead to defects and deviations and incur the cost of poor quality through repairs, replacements, or rework. Thus, a robust mechanism is to be put in place while defining inspection requirements, procedures, and acceptance/rejection criteria, etc.

Practically, customers may not carry out inspections of all lots and all materials. However, defining the process controls, procedures and adherence to the same can significantly prevent the material rejections, damages, and defects. Depending upon the category, complexity and risks involved, the following stage inspections can be decided by customers:-

a. **Pre-production inspection** – This helps to control the quality of inward material at manufacturing works or stores. This is also known as "First Stage," "First Article" or "Raw Material" inspection. This inspection method is used from the primitive stages, wherein the raw materials, components or parts of the final equipment are thoroughly inspected before the start of manufacturing of the main equipment. Its prime purpose is to give objective evidence that all engineering, design, specification, make and quality requirements are correctly understood, witnessed, verified, and recorded. During inspections, the customer (or its inspection agency) can verify the raw material/components and witness relevant tests, check reports and laboratory certificates to ensure that the raw material being used is of suitable quality and standard. Upon approval of this inspection, the next activity starts with manufacturing or production.

b. **Manufacturing inspection** – This is carried out at manufacturing works or shop-floors. This is also known as "During Production Inspection" "Stage Inspection" or "Intermediate Production Inspection." Such inspections are carried out during manufacturing, while the material is under production to ensure that the material being produced in conformance to the agreed technical specifications, standards and manufacturing processes. Any defects, deficiencies or deviations observed during inspection can be rectified at the manufacturing place and defect rate for balance production lots are eliminated or minimised. Such inspections may be carried out on sample material (or proto) before the mass production starts. In case proto is not possible and the main material is rejected (and repair is also not possible at the manufacturing works), it becomes a big risk for the project, as it requires re-manufacturing, which has cost and time implications. In certain cases of defects, the customer may ask to carry out detailed root-cause-analysis (RCA), to ascertain proper repair or rectification of defects, as it may become very late if the final or finish goods are ready, especially when there is no or low chances of modification or rectification.

c. **Production inspection** – This is commonly known as factory acceptance test (FAT). Such inspections are undertaken when manufacturing of the material is completed and placed at finished-goods zone or stores. Customer or its representative can physically witness and inspect the material. If required, the customer may perform agreed functional, routine or any other tests in the factory. In case of minor issues, defects or performance, the customer may advise appropriately (repeat test or rectify the defects), before shipment. Parties must agree in advance about the test to be performed during FAT. Sometimes, materials may be required to undergo specific type tests or functional tests at third-party independent laboratories. In such cases, the parties should expressly agree on the time and costs involved in carrying out such tests at external laboratories. Customer inspection outcome may include acceptance, conditional acceptance, or rejection of the material. If the final inspection of production is approved, the material can move to the next stage i.e., the pre-shipment inspection stage. However, at times, materials are shipped directly from external laboratories to the project sites. In such a case, necessary logistics should be properly taken care of as per agreed norms.

d. **Pre-shipment inspection** – This is taken up before the shipment of the material from the manufacturing unit. This to ensure that the quantity, quality, and packaging of the material are intact, so that the materials reach safely at the destination, although all materials require good packaging so as to withstand the jolts and jerks during transit. However, special care needs to be taken for the shipment of fragile items like ceramic, glasses, lighting, solar panels, etc. which are the costliest and carry the highest transit risks. Many manufacturers define clear guidelines and standards for packaging also. Poor packaging can lead to multi-dimensional movements of goods during transit and cause damages, which can only be detected at the site while opening the packaging. Risks related to poor packaging, incorrect transportation or falling from moving vehicles could have devastating consequences.

e. **Loading & unloading inspection** – Material loading in lorries, containers, rails, aeroplanes, or vessels at origin and shifting at intermediate locations or unloading at site locations are crucial steps in the safe delivery of the goods upto destination. If a good-condition product is incorrectly loaded or unloaded, it can damage the material. Thus, relevant instructions or guidelines of manufacturers must be adhered to and documented properly. Negligence, undue pressure or use of under-capacity machines (cranes, forklifts, hydra, etc.) during loading and unloading may risk material, humans and other properties. The contractor should plan in such a way that heavy equipment can be directly unloaded on foundations or plinths, like large transformers, panels, inverters, gensets, turbines, etc., rather than unloading elsewhere on the ground or in the stores. Wherever possible, indoor panels and material should reach the site upon readiness of the indoor areas (like buildings, control room, etc.) to avoid unnecessary environment exposures and other perils.

f. **Site Acceptance Inspection** – This method involves the visual and physical inspection of the material upon arrival at construction sites. These may involve visual witnessing of shipments on loading in lorries, unloading at the site, shifting at site stores, or working fronts. Many contracts or situations may require photographs, videos, or joint inspections of material at site acceptances. In case of heavy, fragile or critical items, sometimes, third-party agencies, or insurance agency may be involved to witness unloading at site. Proper documentation and acknowledgement of the goods must be made upon site acceptance inspections. Due to packaging, as the goods may not be visible, conditional acceptances can be accorded, subject to the opening of the packaging, as some inherent defects, cracks or damages could be detected later. Contractors may take appropriate insurance covers related to concealed damages.

As such, there are multiple stages, when the material can be or required to be inspected and every stage carries a risk of rejection of material due to non-conformance to technical specifications or quality procedures. Albeit, the inherent or latent defects may still not be detected immediately.

Parties must clearly agree on the Incoterms and mode of cargo movements, as the different modes of transport (road, rail, air, water, or multi-cargo) pose different costs, times and risks. For international trades, parties should expressly agree about custom payments, customs clearances, exemptions applicable, and planning for local movements, insurances, etc.

SCM and logistics team should properly understand and adhere to the government compliances, foreign trade policies and foreign trade agreements among the countries to take benefit of related exemptions, benefits and incentives of manufactured products. Such benefits may include duty-drawback, MEIS, SEIS, EPCG and advance authorisations, duty-free import authorisations etc.

5. Compliances & Ethics Risks

Corruption, distortions, scams, and frauds are significant threats and common risks to projects and businesses. This includes activities like conspiracy, collusion, conflict of interest, bribery, misappropriation, manipulation, misreporting, etc. There is a higher probability of ethical risks, where authorities or controls or powers are conferred to internal employees, outside consultants or other stakeholders. Moreover, wherever administrative powers or financial authorities are provided, the risk of ethics and compliances may also increase.

All organisations prefer to work with suppliers or partners who strictly adhere to high standards of compliances and ethical practices. These are the foundations to build long-lasting trustworthy relationships among stakeholders. Rather, great organisations start their negotiations by executing proper undertakings or assurances of compliance so that any such unethical or influencing attempts are not made by either party to defeat the core objective of the project or business.

While evaluating or assessing your suppliers, contractors, or other agencies, the customer should also ensure that their employees, partners and all business associates adhere and honour the compliance requirements like ethical, legal, regulatory, financial, or commercial – be it internal or external. Eventually, compliances and regulations must be part of an organisation's strategy. For example, all contracting organisations are subject to compliances with prevailing labour and construction laws. Partner organisation must implement the compliance management solutions by setting up the suitable processes, policies, procedures, programmes, audits and reports, etc. in accordance with the relevant laws and regulations.

It is easier said than done. All organisations may define or propagate strong ethical values or codes of conduct. But when it comes to implementations at the ground, they fail to clearly communicate down the line, or implement effective checks, balances and controls. It cannot be implemented by merely defining the policies, but need to evolve and develop the culture, where core values should be non-negotiable for all, top to bottom, in the organisation. In the event of any such violations, the organisations must take serious and stern actions promptly and set the example for the rest of the employees.

Ethics is a science of morals. This removes all types of discrimination and boosts mutual respect for all. This may also require them to execute agreements on code of conduct, non-disclosure agreements or confidentiality agreements also before starting a dialogue for the project. Any violation thereof may be treated seriously and appropriate disciplinary actions taken against the accused person.

6. Monopoly Risks

There may be exceptional situations when the customer organisation is dependent on an exclusive, proprietary, or non-substitutable supplier in the market. This is common while a specific design or patented technology is required by the customer. The major risk with such a monopoly supplier is that the customer has very less scope of negotiations and is steered by the supplier in their benefits.

One of the possible ways to mitigate monopoly suppliers is to look for design, patents or technological solutions which can be offered by multiple suppliers or manufacturers and are also backed with a wider presence in market at multiple locations, factories, authorised channel partners or distributors, etc. This stimulates their confidence so as to de-risk the monopoly or concentration on a single supplier and ensure that it meets the requirements and quality deliverables with minimum hassle. However, if the contractor has not envisaged the monopoly risk of sourcing, then it will have a significant impact on their plan and budgets.

7. Logistic Risks

Logistic risks are those related to the overall logistics operations. This covers the risks encountered at any stage from manufacturing, inspection, packaging, dispatches, distribution, transportation, storages, compliance, etc. Logistic risks can occur due to multiple reasons including negligence, mishandling, improper packaging, chaotic warehouses, weak documentation, compliances, socio-political disturbances and many other human pirated or natural exposures.

A project normally encounters the following types of logistic risks:

- Delay or shortage of raw material
- Delay in the availability of vehicles, rails, vessels, or containers
- Delay in clearances at ports, airports or custom houses
- Inadequate production capacity or allotted capacity of key suppliers
- Shortages of workers, frequent strikes, or industrial unrest
- Non-standard or poor packaging issues
- Incoterms, risk of loss, risk of title-related issues
- Non-acceptance, non-adherence, or unfair contract terms
- Inclement weather or unfavourable sailing conditions
- Technical uncertainty, rapid developments, and certifications
- Weak compliance to quality and HSE standards
- Obsolescence of products or suppliers' disappearance
- National or global trade regulations and cross-country disputes and policies, etc.

Logistics in projects have a crucial role to play. Logistics mandate a systematic approach in managing the flow of resources (equipment, material, machines, assets, tools and spares) from the source to destination. If project logistics are not managed well, project efficiency can turn into chaos, in no time. The role and responsibility of the logistics team are actually much higher and wider than understood by most people. This may include planning and controlling of purchases, manufacturing, inspection, packaging, loading, unloading, transporting, insurance, custom clearances, stocks & inventories, security, safe delivery in good conditions and documentation.

At times, the proximity of factories plays a vital role in mitigating the risks of delays or ensure just-in-time supplies of material at sites. The criteria include the availability of factories closer to construction sites. This is more preferred where requirement or frequency of purchase is high. For example, steel, cement, other construction materials, structural steel, etc. are used in high volumes in the construction project and parties prefer the availability of the same in the closest vicinity or source of supplies.

Logistics has its own typical deterrents and constraints:

- **On-Safety** – Material handled, packaged, shifted, stored and used in the safest manner from origin to destination without any harm or damage to material, person or third party.

- **On-Time** – Deliver the material right on time – neither too early nor too late, as the project can neither afford excessive pile-ups of inventories nor abnormal delays.

- **On-Place** – Deliver the material at the right place, using the shortest and safest route, without missing or mixing with other material or projects.

- **On-Quality** – Despite using multi-model cargos, the material must withstand and sustain all jolts and jerks and remain intact and perfect during transit.

- **On-Cost** – Deliver within the right amount of the budget and cost. This is possible with various combinations and permutations through different models, routes, and feasibilities.

Handling of break-bulk cargos, especially those required for heavy engineering industries or plants needs special expertise, care, and protection from the origin to the destination. Logistics play a most vital role in handing such break-bulk shipments. Likewise, if a project involves specialised chemical, fertilisers or fragile items, special expertise and packaging are crucial, because these are susceptible to flames and fire, which can cause major incidents and accidents.

Nonetheless, logistics damage to the material may have a heavy impact on project cost, schedules and might shift project deliverables.

In order to eliminate or minimise the logistics risks in projects, the following preventive measures can be adopted: –

a. **Inbound & outbound logistics** – Supply the material only when actually required at sites and not in much advance. It means material ordering, manufacturing, and shipments are to be closely integrated with project execution schedule of sites so that the material reaches site when fronts or foundations or buildings are ready to accept the same.

b. **Production logistics** – Raw material purchases, manufacturing, assembly, fabrication, packaging, etc. are so planned that material reaches at the site on time; it incurs neither unnecessary idle charges nor undue piles of inventory. Very early receipt of material or too big piling of inventory can have adverse repercussions on the project cash-flow, storage, insurance and warranties.

c. **Reverse logistics** – Many materials or equipment may require to be returned from sites to manufacturers' works for various purposes. The logistics team also ensures to establish such processes so that rejected, surplus or serviceable material is sent back with clear logistics in the most optimal manner.

d. **Emergency logistics** – In the event of a critical breakdown of any equipment or material or plant, how quickly the spares or components can reach the destination is important so that the defects can be quickly rectified and normalcy restored. This can directly benefit and contribute to minimizing the downtime of the equipment, machines or plants.

Eventually, the supply chain environment is prone to high risks and vulnerability. With the increased demands, the complexity of trades and logistics is growing rapidly globally. Despite posing enormous challenges and exposures, logistics risks are less evaluated and very few companies genuinely reflect the logistics risk in their risk register.

8. Storages & Reconciliations

Most construction projects carry high risks and exposure of mixing, missing, thefts, damages, and pilferages at sites. This not only burdens the project with additional costs of replacements but causes unnecessary probing, investigations, insurances, documentation, delays and administrative interventions.

Commonly, it is the responsibility of contractors to ensure care, custody, protection, security and storage of the material, equipment and plants. Thus, all the material received at site, whether purchased by a contractor or issued by customer, should be properly unloaded, receipted, stored, stacked, and documented so that it can easily be reconciled at any time. Thus, a robust inventory control mechanism of receipts, issuances, consumptions, and reconciliation should be put in place.

As contractors are responsible for inventory management, following precautions are to be taken while bringing the material at the site:

a. Provide complete details to sites in advance related to material, Bill of Quantities, specifications, sizes, volumes, quantities and approved vendor lists.

b. Prepare standard operating procedure (SOP) for material receipts, issuance and storage with proper marking and stacking, at earmarked places.

c. Material receipt at site must be tracked and shall conform to details (as in point a above).

d. Material inspection procedure, process and place (factory or sites) should be well-defined.

e. Provide relevant documents (inspection reports, dispatch authorisations, packing list, invoices, lorry receipts, bill of lading, insurance covers and others), with each shipment.

f. All inward materials are inspected by authorised persons (construction, stores, quality or commercial) before accepting and entering into stock registers. Proper goods receipt notes (GRNs) or material receipt notes (MRNs) are to be made as early as possible.

g. Heavy or fragile equipment may require joint inspections/photographs be taken on-lorry (or on-container) conditions before the start of unloading.

h. Any damages, defects, short-supplies, extra or surplus quantities must be reported to the concerned authority immediately. In case there is need to intimate the insurance company for claims, the same must be reported immediately, in consultation with your insurance, finance or legal department.

i. Damaged or defective material must be kept aside and not mixed with good material.

j. No defective or damaged material may be issued for work.

k. Only material in good conditions should be issued for erection or consumption.

l. Complete inventory stocks must be reconciled frequently to avoid futuristic gaps or mismatches.

Many items require a unique controlled serial number; records should reflect these numbers for easy traceability, accounting and reconciliation. Prompt traceability is crucial during warranty also; else its warranty may stand void.

Many project sites do not reconcile the inventory frequently and then land up in a mess or conflicting situations at the end. This becomes more difficult when the staff members change or quit.

Contracts, Bill of Quantity or billing break-ups must clearly define the exact material details that shall be supplied to the project sites so that the construction team can properly track, record, and maintain the same methodically.

Mostly damages or shortages are recovered from contractors. Hence, in order to avoid any unnecessary back-charges or recoveries, proper stocks must be maintained. All projects require material reconciliation of Orders, Bill of Material, MDCC, GRNs/MRNs, and consumptions. Defective, wastage, scrap, damaged, or unused material should also be reconciled properly mutually between the parties.

Nonetheless, the biggest ambiguity or conflict arises from the engineering or running items like earthing materials, cables, conductors, wires, lattice structure materials, lighting material steel and cement, carpenters' or plumbing material, etc. As such items are considered in lots or lump-sum, it becomes difficult to reconcile if the records are not well-maintained from the beginning. Further, special care is also required to handle the fragile material, like solar panels or glass or ceramic items, which pose the highest risks and highest cost.

9. Insurances & Claims

Insurance requirements in construction projects are complex and generally, all risks and responsibility of the same are included in the contractor's scope and associated costs are assumed in the contract price itself. In order to mitigate such risks, contractors procure various insurance covers. Common insurance covers include marine risks, transit risks, erection-all-risk (EAR), construction-all-risks (CAR), automobile, civil general liabilities, etc. Typically, these insurance policies cover the risks to the extent of the reinstated values of the supplies or work plus a mark-up towards associated costs related to consultancy and professional services that may incur. Thus, keeping the gross perils and insurable interest in mind, the contractor must negotiate a good insurance cover for the project.

Insurance is a subject matter of solicitation. This is a very tricky tool or punchline. It means that no standard insurance completely covers all risks. Thus, all types of potential risks, defects and damages are to be discussed well, negotiated and documented with insurance companies, and thoroughly negotiated before concluding the insurance policies. Insurance policies must define the full scope and work carefully. An insurance clause in the contract defines the insurance requirements and type of insurance policies that parties need to procure and maintain throughout the project lifecycle. Some contracting parties may also agree on additional terms, add-ons and endorsements which are to be included in the insurance policies. Often, contractors pay the premium to purchase the insurance policies but at the real time of need (or claims), they may find that policies impose many restrictions and exclusions, which make the claims either partially acceptable or may get rejected outright.

Most construction contracts require the contractor to indemnify customer from and against any and all losses, damages, costs, charges, fees, claims and/or proceedings arising out of or with respect to:-

a. Use of allocated land and site infrastructure

b. Death of or injury of any person, including a third party

c. Loss of or damage to any property, including a third party

d. Remedying any defects arising during execution and completion project

Thus, the insurance clauses impose that the contractor has understood and satisfied the contract requirements and ensure the guarantee of protecting parties' interest. Any negligence in implementing these requirements may lead to serious consequences and breach of contract, if not adhered to. Parties to ensure that:-

- Fully understand the scope, responsibility, obligations, perils and risks to ascertain its insurable interest.

- Insurance requirements are reviewed by domain experts and confirmed with market norms and insurance covers available in the market.

- Suitable insurance policies are available and premium costs are admissible.

- Insurance adequately covers full scope and obligations. Parties expressly know the inclusions, exclusions and endorsement in the policies.

Many contracts cap the liability of the contractors or suppliers, to the extent of the contract value, with certain exclusions. However, in case of damages or claims, this understanding becomes a point of major conflict, as contractors are liable to the customer for the total scope of the contract. There may be differences in understanding. Many customers freely issue heavy equipment to the contractors for erection, testing, and commissioning purposes. Contractors are thus responsible for its complete care, custody, production, storage, handling, erection, and commissioning. Any damages to such customer-supplied equipment may be recoverable from the contractors. Hence, it should be aptly covered in the insurance policy, as contractors are obliged to handle and complete the scope with customer-issued material also.

Interestingly, it is frequently experienced that many such insurance claims, where customer-issued materials were not included in the insurance policy and led to major losses, as the insurance companies decline to entertain the claims or compensate the losses.

There are instances where contractors agree on lump-sum turnkey contract prices. Later, the parties mutually agree for a billing break-up (BBU) and the contractor raises claims according to the BBU. However, this BBU may not reflect complete items or obligations, which the contractor needs to fulfil. In case of any damage or defect in the items of work, which is not covered in the BBU, the contractor is liable to replace or repair the same, whether it is included in the BBU, insurance cover or not.

Organisations may procure different policies from different underwriters to cover different stage-risks. However, at the time of the loss, none may easily accept the claim, as they tend to prove that the damage occurred is not in their jurisdiction and avoid stating that such damages fall under unsubscribed stages, events, conditions or policy. Therefore, contractors must ensure that there is a seamless integration among all the policies. Such cases are normal when the material gets damaged. For example, if the material is found damaged during the opening of the packages, the EAR insurance policy may deny the claims stating that these damages occurred during transit and are to be claimed under transit insurance, as such damages should have been detected while taking possession of the material. But it is not possible to open the actual packaging at the site while receiving the material in big lots and volumes. In such a case, the contractor may struggle between different policies. So proper covers and add-ons may be considered related to concealment of damages, where material packaging is found open after a certain gap and not immediately upon receipt.

Similarly, a common scenario of conflict is seen that plants are commissioned and put on service, but contractors still have to perform the balance works and the projects are not handed over back to the customers. As such, the contractors are liable to hand over the plants in good condition to the customer before the provisional or final acceptance milestone is achieved. Upon commissioning of the plants, normally operational insurances are triggered, which may otherwise cause conflict between EAR and operational insurance cover. Such issues may be handled very carefully by the insurance experts so that in case of claims, there is no dispute and claims are settled swiftly.

Thus, it is strongly advisable to take expert advice while procuring the insurance policies and ensuring that complete risks and perils are properly covered in the policy with seamless integration between different policies.

Timely identification and assessment of the supply-chain-related risks enable organisations to plan effective actions or mitigations to those risks. Several tools, techniques and strategies can be applied by the team to reduce or minimise the impact of such risks. This may include changing the sourcing plans, exploring different vendors, locations or timings of ordering, splitting or distributing orders among multiple vendors, etc.

Chapter 7

QUALITY RISKS

"Quality is never an accident. It is always the result of the intelligent efforts"

– John Ruskin

Defects at construction sites have always been big risks and puzzles for construction professionals. These defects are one of the main causes of disputes, discrepancies, conflicts, and delays in projects. Construction defects can be caused due to various reasons related to material, workmanship, design, systems, processes, changes, or other subjectivity.

Quality related risks can broadly be categorised as under:

1. Equipment and Material Risks

All contracts require that the contractors or suppliers must use new, unused, and high-quality material, as such materials are expected to last longer, should not have any defect, and perform smoothly for its purpose and design. However, construction projects may still receive materials which are defective as a result of sub-standard, inferior, or poor raw material used or poor manufacturing processes followed by the manufacturers.

Inferior-quality material used in building construction works can cause problems of leakages, cracks, collapsing, falling, or malfunctioning or unstable structures. Reasons for such defects or damages could be many – intentional, unintentional, known, or unknown. Some building material suppliers or equipment manufacturers may use poor raw material, poor manufacturing processes or casualness in workmanship during manufacturing. As a result, such products may not conform to the agreed specifications and standards, or withstand tests. It is also possible that such defects may get missed, neglected, ignored, or compromised during inspections. All such material defects can cause damages to the project—sooner or later. Although some latent defects may not be visible immediately, the same will be encountered at a later stage and may have expensive consequences.

Poor quality or defective material may also lead to accidents and injuries. Electrical equipment defects can result in mild-to-fatal casualties, such as electrocution, fire, explosions, burns and electric shock and other potential hazards. In some cases, electrical equipment defects can even result in major casualties, including deaths. Defects in large scale equipment and machines can turn into deadly situations for workers and operations, due to faults of none of them. Accidents involving defective equipment and machines could have catastrophic consequences, hence such risks must not be ignored.

Contractors must ensure that heavy equipment and machines used (temporary or permanent) at construction sites or production units must strictly adhere to the technical specifications, safety requirements and quality standards.

Construction equipment used at sites like trolleys, cranes, forklifts, JCBs, crushers, shredders, excavators, bulldozers, wheel loaders, transit mixer etc. should be of adequate capacity, strength and with valid calibrations. These equipment and machines should be operated by authorised and skilled persons only.

Delivery or use of defective and damaged material is a major risk to the projects. First and foremost, the contractors must put strong inward material processes in place, to ensure that the received materials are in good condition. Petty contractors sometimes use old and shattered tools & tackles, which are not properly segregated, exposed, shielded, or taped, especially, the construction electrical machines and gadgets which have high exposures due to oversight or negligence of human workers or visitors. Also some contractors can use the old or left over material from other project sites, which are unsafe and unhealthy for the projects.

Contractors should establish a suitable process controls to ensure that the design, construction, materials and workmanship are to be in accordance with contracts, specifications, standards and good industry practices.

Customers and contractors put in place proper controls before materials are shipped or accepted at the site. Controls can range from simple to severe. This may include the issuance of a simple "Certificate of Compliance" or "Certificate of Conformity" based on initial screening or random inspections of material to 100% stringent inspection or testing of materials. Regardless of the controls adopted or exercised by an organisation, it must be remembered that the controls have to ensure 100% conformity for the complete scope or matcrial and not limited to the sample chosen. Thus, quality inspectors or auditors must establish that they stand accountable for the full scope or volume being procured or approved and not for a selected few. Although in some cases, it may not be practically possible to inspect the full volume of material by quality inspectors, however, process quality controls should be established in such a way that at every stage of inspection, defects or defective material could not pass through or pushed to the next level. To the extent possible, defects to be prevented from occurrence or detect at the source and not passed on to the subsequent stage, thus process should be mistake-proof.

2. Service Risks

There is a common saying in projects that installation details should not be designed to be understood, rather they should be designed to be not misunderstood.

Construction defects are common in projects and can occur due to various reasons. The majority of defects identified by field teams may be minor or inconsequential. However, the major defects can have devastating consequences. Such major defects could be a direct threat to the safety, operation and reliability of the plants. Irrespective of reason or cause of defects – the problem remains. Mostly latent service defects are not discovered easily and quickly.

Upon identification of defects, if the same is not documented and rectified immediately, it has high probability of getting ignored, hidden, or neglected. This may cause undue conflicts and discrepancies among parties. At times, such defects may not be taken seriously during the construction phase, but at the time of testing or operations, such gaps or defects crop up and become a major concern for the organisation. As such the rectification of some defects could be quite expensive or sometimes even impossible, at a later stage. For example, if one of the beam in any building is tilted and has cold-joint, then it would not be feasible to carry out the 100% repair or rectification and customer may have to live with that only.

In order to avoid or minimise such defects, parties must agree upon the clear and detailed field quality plans (FQP), quality procedures and acceptance criteria to resolve any non-conformity. There must be a clear-cut process defined wherein parties duly witness, record, rectify and close the gaps in a structured and documented manner. Many good customers enforce that contractor's work measurements or payments are certified based on clearances for their field quality and field safety teams. In case of any unresolved defects, the work certifications or contractor payments shall be directly impacted.

Another risk is procrastinating the rectification of defects for a later stage. This is common in the construction industry practice. While the construction team's primary focus is to achieve higher productivity, they tend to ignore or defer the minor defects or issues, which get grossly neglected or become major risks for the project. So, the team should work in such a way that any defect, if arises, should be rectified immediately and properly documented with its proper identification and closure reports.

Due to vast size, areas, scopes and volumes of large projects, thorough inspections by field quality teams may not be feasible at every stage. However, quality process controls and audit procedures should be established in such a way that quality team could draw relevant inferences from the samples and drive the actions thereof. Nonetheless, construction quality is not the exclusive responsibility of quality inspectors, but the entire construction team, who constructs, supervises or monitors the project work.

3. Inadequate Supervision

All contractors must ensure that every task, activity and material used at the site must be as per approved specifications, field quality plan and agreed processes only. Contractors must ensure that all work fronts are properly supervised and monitored by skilled and competent supervisors, all the time. It starts from marking the land contours, layouts, profiling, area grading, excavation, PCC, raft, cube works, slab works, piling or foundation works, roads, drainage, culverts, or all other equipment erection works, etc. Enhanced supervision or control is required while the construction teams work at elevated platforms, heights, deep excavations, complex structures, during night shifts or extreme weather conditions etc.

Commonly, contractors or their sub-contractors do not give attention to detailed planning, budget or deploy adequate quality resources during the construction phase. As a result, they not only lose the opportunity to detect the defects at the early stage but also lead to reworks and higher cost of rectifications later. So, adequate quality team with engineers and supervisors possessing adequate skill, experience, and competency should be deployed at the site. Contractors can plan for in-house or third-party agencies for quality inspection of works.

In the absence of proper supervision from a quality and safety perspective, there is a high probability that the construction sites may invite disaster, damages or casualties without knowing it in advance.

4. Zero-Punch-Point

In the construction industry, every single front has multiple opportunities for defects, errors and mistakes. So very clear focus and emphasis must be given to each and every task during construction, so as to avoid large numbers in the gap-list or defect list, during completion stage.

All projects are temporary and at some point in time, they end. Before projects are put in service or handed over to the customer, the parties jointly carry out inspection or witnessing of the punch-list points. Identification and closure of punch-list points or snag-list points are widespread and a traditional practice in the construction industry. These punch-lists cover all the balance or pending works, damages, defects, deficiencies, or aesthetic look of all completed works. Generally, a list is prepared for the tasks which are in non-conformity to the contract, specification, drawings, or prudent industry practices. The contractor is then liable to correct the punch-list items before handing over.

If there are major punch-list points, they can become a threat to the safety or operations of the plants or equipment. In such events, further operations may be suspended or held up until such critical or major defects are rectified, the root cause is ascertained and corrected, including modification of processes, documentation, and check-sheets so that the defect is not repeated at the site.

In large plants, managing the punch-points becomes a project by itself and can consume more than 20% of the total project duration, as the identified punch-points can run into large numbers and closure will take much longer time and resources. This not only impacts the plant performance but leads to time and cost-overrun for contractors and customers. It also affects the opportunity costs by not shifting the existing resources to other projects. It is common that most of the petty sub-contractors complete their respective scope of work and leave the project sites. If the punch-points are identified at the end of the overall project, the petty contractor would have already demobilised its manpower and resources. Thus, it becomes all the more difficult to get such defects rectified. It is, therefore, strongly recommended that wherever possible punch-point identification and defect resolution should commence in parallel, when the core construction work is in progress, rather than grossly delaying until project closure stage.

Generally, punch-points can be categorised into two: (a) Major, and (b) Minor. Major punch-points are generally those which are major concerns or pose high risks for the operations, safety or strength of the equipment or plant, the rectification of major damages and major balance works, whereas all other punch-points are categorised as minor.

This is one big grey-box-opportunity in the construction sector, which needs to be tapped by all organisations in construction. If HSE aspirants can target zero accidents, quality aspirants target zero defects, why not construction team target to achieve zero punch-points at sites?

Achieving zero punch-points may look an absurd idea to some professionals, but that is where the future lies for the construction industry, as time and cost are always big constraints in projects. Albeit, achieving this level is neither easy nor technically possible in the perfect literal standpoint at all works. However, all error-proof processes, controls and efforts should be made in the project from inception so that punch-list points can be brought to a negligible level or achieve first-time right construction.

A good practice is that when a contractor completes any activity, it should immediately identify defects or punch-lists and rectify the same immediately before moving to the next or succeeding activity. At this stage, sub-contractors, workers and tools & tackles are available for rectification. Only after the resolution of the punch-list points, the next activity should start. For example, if you are starting the final painting in a room, you must ensure that all the plumbing, sanitary or electrical internal workers are completed and any gaps or breaks or damages to the walls are duly repaired before painting is carried out.

Handing-over phase is such a phase of the project while all hard, complex, and difficult actions taken by construction teams may generally be ignored, as the prime focus is only identifying gaps, defects, and errors. If this punch-point process is not managed well, the organisation may end up making multiple punch-point lists and the closure becomes a never-ending chapter.

To achieve zero punch points, the key initiative may include:

1. Believing, spreading awareness and culture about this mission is critical

2. Agree on acceptable standards, inspection and audit criteria from the start

3. Fix sources of material for better constructability and eliminate the possibility of defect.

4. Doing the right things, right the first time and every time right.

5. Fixing every mistake at source and not on the specific spot only.

6. Stage gate quality clearance at all critical construction processesSampling alone may not be a winning strategy. It may not guarantee zero defects; so controls and inspection plans should cover the entire scope.

7. More inspectors only increase defect detection and reworks, so engage all partners in developing quality culture and quality excellence programmes.

8. Every work is carried out with a vision to avoid 'cost of poor quality' later. Prevention is less costly than failure. (1:10:100 rule).

9. Quality function should be independent with accountability and without undue influence, bias and reporting constraints.

10. Process redesign to eliminate mistakes by applying PFMEA in all construction processes to anticipate and mitigate risks. The Quality Programme must address the mitigation of all defects which are likely to happen during construction.

Enforcement of the above guidelines can bring down the cost and time significantly.

Parties should agree on the quality control mechanism to ensure compliances with the contract provisions like quality assurance plan (QAP) and field quality plan (FQP). These quality control mechanisms should expressly cover the followings:-

a. Establishing the accountability of quality deliverables or defining quality organisation, duties, responsibility and authority of team members

b. Procedures, inspections, audits, and documentation

c. Testing of material (Testing agency, sampling methods, testing frequencies, testing places, testing facilities, applicable standards, acceptance and approval criteria, recording, reporting, etc.)

d. Costs related to resources, tools, instruments, prototypes, samples, manpower required to conduct these tests.

e. Quality audit system – procedures, checklists, frequencies, reports etc.

f. Procedures for rejection, remedial actions,

Quality team should emphasise on quality process during construction and avoid rear-facing inspections, rather look at upstream processes.

5. Quality Culture & Leadership Commitment

Quality is not the exclusive responsibility of any individual or function, not even of the quality team. It is a collective responsibility of the complete project team and organisation. This requires a strong commitment from the leadership, with a top-down approach.

If you are engaging with down-line vendors or suppliers, the entire organisation should exhibit positive behaviour towards quality culture and the organisation structure should reflect the same. It needs to be duly assessed if their staff demonstrates the quality in their processes, communication, behaviour, meetings, commitments and documentation. Is the supplier or contractor organisation being open, fair, transparent, and cooperative – with internal and external stakeholders?

Most important is the involvement and commitments demonstrated by the leadership of the supplier or contracting organisation towards implementing quality controls and their willingness to accept feedback or reaction to handle any issues or defects. This is possible through their inquisitiveness towards quality, and the same is visible through their personal interactions, reviews, communications, visits to shop-floor, project sites, customer locations and actions undertaken by the leadership.

6. Quality System and Procedures

Another key reason for quality risk is the lack of enforcing quality system or governance at construction sites. For example, suppose it is designed to use "M20 concrete mix" and must be approved by a competent quality manager. However, if the same is either not checked or checked by non-competent personnel, there is a high probability that the concrete poured to the foundation may not match the approved design-mix for the foundation. This can become a major risk for the strength of the foundation or structures. So, strong quality systems must be put in place to ensure no compromises are done and proper mix or supervision is available.

7. Drawings & Document Risks

At every construction site, the most common risk contractors or sub-contractors encounter is related to gaps in drawings and actual construction. These gaps could emanate due to various reasons including geometrical differences between AutoCAD drawings, constructions drawings, multiple versions, etc. At times, customers take a long time to approve the good-for-construction (GFC) drawing or respond to the clarifications; as a result field teams are either forced to suspend the site works or carry on with construction by taking risks under the pressure of large resources and construction material. But in between this hustle, field execution is disrupted, or idle labour costs are incurred.

All construction work is to be carried out as per drawings, specification and documents provided by architects, design, or technical team to the field constructions teams. Rather, a robust system must be put in place between the technical team and site construction team so that the flow of drawings and technical support is streamlined and in case of any information or clarifications required by the site team, the same must reach without much lapse of time.

Every construction organisation should adhere to a certain quality system, document it and get it approved by authorised leaders. This provides a proper way and guidance to ensure that all details of work fall in place. There can be a large number of procedures, which are used in construction sites, which can include auditing, procurements, inspections, material approvals, non-conformities management, reporting, controls, workforce training, problem-solving, etc. Proper usage of such procedures can largely reduce defect levels.

Chapter 8

EXECUTION RISKS

"Those who plan do better than those who do not plan, even though they rarely stick to their plan"

– Winston Churchill.

If you fail to plan, you plan to fail. Having a plan in place and working according to it increases the probability of success. Though the project plans and schedules keep rolling and changing during the construction phase, having no plans or schedules in the project is the biggest risk. Some project experts may execute the project activities based on their mental assessment, guesstimates, and experiences which is not only an individualistic approach but also a poor management style, as non-documented plans run the risk of incorrect estimation or missing many tasks, activities, and dependencies. Absence of a project schedule is the risk and should be documented in the risk register. Subsequent progress monitoring and variances should be in accordance with the baseline plans.

Construction projects involve a large number of issues, complexities, and variables, and often it is very difficult to ascertain the actual causes and consequences, dependencies, and correlations. Hence, risks play a vital role in decision-making and project execution.

There is a very powerful six-P formula of *"Proper Prior Planning Prevents Poor Performance."* This can contribute to increase productivity, performance, and probability of success. Though plans may be useless, planning is indispensable.

Execution risks can significantly impede the project activities and procrastinate to achieve the objective on time and cost. Sooner or later, most construction projects are subjected to overruns (time and costs) and lead to tremendous pressures on time, cost, quality and resources throughout its project life cycle ("PLC"). One of the main reasons is the over-optimistic planning.

At the beginning of an opportunity, most organisations give high priority to the evaluation of financial and legal risks whereas execution risks are overlooked. But execution risks are equally crucial in projects and start from inception while an organisation participates in a tender or clinching a deal with a vision to accomplish the complete scope or goal in the specified time overcoming all risks, constraints and challenges.

Identification of execution risks also starts from defining the scope, conducting site surveys, associated infrastructure and deputing environmental experts to determine the social-environmental feasibility of the project. The contractor should carry out the comprehensive analysis of the feasibility of executing the project. This includes a clear understanding of the scope and deliverables, undertaking initial surveys to assess the ground complexity and impediments of site, access and surroundings which can jeopardise the actual construction if ignored or incorrectly assumed.

It is a myth that by deploying the best resources of the organisation, risks can be brought to zero. Execution risk management is a continuous process during the construction PLC, though some latent risks may be detected later during operations.

For example, the world-renowned Boston Big Dig Tunnel is a classic case of all the risks faced by any large project. This project was mammoth, technically complex, operationally challenged, commercially uncontrolled and experienced the highest levels of uncertainties, obstacles, and difficulties. The project cost shot up from the original budget of US$ 2.80 billion to US$ 14.78 billion. Fluctuations, inflation, and claims played a significant role in this project. On the schedule front, the project took almost 25 years from concept to construction. The project was envisaged to resolve the traffic gridlock situation in the centre of Boston, USA. The project faced the key challenges of (a) Scope changes and extra claims, (b) Price inflations of material, commodity, interests and labours, (c) Weak project management and (d) Design and technical complexities, ('e) *Force majeure* and unanticipated site operational risks, (f) Legal and environmental risks, etc.

The following execution risks are very common in construction sites, which every project or contractor faces with less or more gravity:-

1. Schedule Risks

Schedule risk means that project activities, tasks or milestones can take a longer time than planned or anticipated. Delay in project schedules not only increases the risk of additional cost burden but also brings additional liabilities and exposures, including liquidated damages, loss of revenues, production, reputation or even loss of purpose. Thus, poor scheduling and planning can have a devastating impact on execution, budgets, and scope.

A schedule may defend the project activities from chaos and whims but managing the schedule in a project itself is a major challenge. Time is the most valuable and scarce resource in projects. It is always a challenge for the project teams to complete the full scope of work within time by overcoming all issues, changes and risks. Defining a project schedule is of importance because it is an opportunity for the project team to split the full scope in detail and envision the resources (manpower, skills, tools & equipment, etc.), which gives a good amount of clarity related to the feasibility of completing the scope on time and enhance probability of success. Forming a schedule is like laying the foundation of the project's success. The schedule encompasses all milestones, activities and tasks to be performed within the available timeframe. If the foundation is not set right, the delivery is unlikely. If scheduling is done well, there are chances that modifications or changes may have less effect during the performance. Further, by implementing a strong monitoring and control system, the variances in schedule can be easily identified, and timely corrective actions can be taken to accelerate or bridge the gaps.

Risks risen from any source may impact the schedule or execution performance. Common risks that may impact executions include, but are not limited to, (i) unclear or improper understanding of scope; (ii) slow approval of drawings, specifications and documents; (iii) frequent changes; (iv) delayed material manufacturing, inspections or logistics; (v) delayed payments or LC openings; (vi) lopsided contract negotiations; (vii) local and political issues at the site; and (viii) shoddy quality of work due to various reasons, etc., Unless professional contract management practices are followed, such reasons continue to impact execution, cause conflicts and sour relationships.

Besides the above, many hidden risks also emerge in projects, which go unnoticed and affect the execution. The project schedule comprises of many simple-to-complex milestones, tasks and activities with various inter-dependencies, which get ignored or slipped due to other priorities. Generally, project schedules are made on backward planning, if the end-date is sacrosanct. So, in case the actual progress is running behind, the organisations can work on the expediting or acceleration plans to recover the time lost.

It is observed that contractors consume a high level of resources and energy at the early stages of the project but the output is relatively slow and as the project reaches near maturity or completion, the work pressure mounts up dramatically. This leads to a lot of compromises or pressure by doing the essential works first and delaying the non-essential ones. So the good plans should be balanced and monitoring the earned values and estimates to completion very closely.

The project schedule must reflect the following key points:-

a. Full scope of the work (including package, phase, milestone, or full project)

b. Current execution plan (as baseline and not future or rolling plans)

c. Correct status of any past task or activity (with actual dates)

d. Customer's deliverables, expectation, and outcomes

e. Correct size and complexity of tasks—not under or overestimated

f. Resources are allocated with each task and activities

g. Define correct relationships and inter-dependencies of each task and activity

h. Buy-in and acceptance of relevant stakeholders

i. Each task and activity should have one predecessor and one successor activity

j. The critical path should be clear and logical

k. Avoid hard constraints that can produce a negative float

There are no global standard templates available about its size and details to prepare project schedules; it depends upon the manageability of the project team or customer requirements. However, project organisations can decide on the repeatability of their projects or past experiences. The project team can detail out the tasks and activities in work breakdown structures (WBS) in the schedule, as much as they can manage. There is no restriction or rules on the number of line-items or WBS items in any schedule. Over-optimistic or unpragmatic planning can have significant changes and consequently impact many other activities.

Projects operate in a VUCCA environment, which can undergo change. The schedule should consider these dynamics of changes. In case of any change in future, the schedule can be revised immediately by clearly defining above parameters, without putting any hard dates or constraints, to the extent possible. Therefore, proper linking of all tasks and activities helps to promptly update the same faster and correctly.

A critical path is the sequence of project network activities which add up to the longest overall duration, whether it has been floated or not. Yet, this determines the shortest time possible to complete the project. This helps the project team to identify the activities which are on the critical path, focus on tasks which require great attention, as delay or failure of any of the critical path tasks can directly affect the project timelines. Therefore, to establish a delay, the contractor must be able to recognise the events that will cause the project to be delayed. A separate delay log can be maintained by contractors right from project start upstage.

The schedule is also important to identify schedule variances through a comparison of baseline schedule and status of the activities on the critical path.

Once the actual project work starts, then all the planning, dependencies, assumptions, and scenarios are put into reality. *If it is not managed well, the baseline will be in* the *base (basement) and* the *actuals will be in space!*

2. Cross-functional Risks

Large projects are carried out at the organisational level, which involves multiple cross-functional members. An amalgamation of multiple cross-functional members is crucial as these members are governed or controlled by their respective functional leaders. Different functions may have their own plans, activities, and priorities, which may conflict with a project schedule. Nonetheless, different functions may also have various constraints related to limited resources and skills. However, in case all relevant functions are not aligned, it can derail the project from its objective. Thus, all cross-functionals must assign, empower, and allocate their suitable resources to invest time on project priorities. In case cross-function team (CFT) members are supporting multiple projects concurrently, their priorities can vary, which needs to be adjusted meticulously with the help of the functional leaders. Therefore, the project team has to ensure a great amount of coherence among all team members and their schedule; thereby, an integrated schedule is agreed.

Many functions play key roles in projects. Key roles and responsibilities performed by CFT involve:-

    a. **Technical & design** – prepare project design, engineering, scopes, specifications, layouts, surveys, technical studies, performance, drawings and other technical documents. Conduct technical feasibility, evaluations of products, services and solutions. Explore new technologies, systems, innovations, and continuous development and optimisations, to enable organisations to implement through projects. Technical experts evolve and experiment new products, features, designs, systems and technologies, faster than the projects can keep pace with them.

b. **Supply chain & logistics** – identify, develop and maintain contractors, suppliers, consultants, and other service providers. Roll-out enquiries, tenders, RFPs, conduct assessments, and negotiations, award contracts or orders, ensure material inspections and deliveries at project sites. Also review and evaluate their performance so that project objectives are not affected. Their role also involves marathon efforts of market intelligence and ensuring compliances to specifications, quality, budgets, delivery, market-level costing and other logistics requirements.

c. **Quality** – the role of the quality team is to prepare and implement a quality management system (QMS), which includes quality planning, quality assurance and quality control. They ensure that projects are executed successfully with the best quality, stated period and minimum budget. With support in finalising manufacturing quality plans (MQP), field quality plans, (FQP) Process Quality Plans (PQP) and other procedures and audit plans, checklists, formats, etc., the quality team ensures that all project materials and services are in full conformance as per contracts, technical specifications, relevant standards, codes, and plans.

d. **Product management** – projects require a large number of products, material and equipment. Product management group ensures that the latest products or material are developed, supplied and used at projects so that they remain abreast with technology, competition, and market demands. They can also ensure that manufacturers have adequate infrastructure and production capacity, so as to plan the execution schedule accordingly. If a single vendor cannot fulfil the total requirement within a limited time period, they can explore or deploy multiple sources or product mix based on their production capacity. Sometimes, organisations may not have a separate team for product management and such activities are carried out by the R&D or SCM team.

e. **Finance** – support in managing project funds, budgeting, costing and cash-flow, profitability and returns. They ensure that all debtors and creditors are managed well without default and other finance and statutory compliances are fulfilled in the time and manner required as per prevalent law or company policies. In case forecasted numbers are impacted or slipping, the same can be immediately alerted by the Finance team so that immediate remedial actions can be taken by the project team or management.

f. Besides the above core functions, other support functions also support projects on need-basis. Such functions may involve Sales, R&D, HSE , Information Technology, Human Resource, Administration, Legal, Risks and Contracts, Regulatory, etc. A good organisation ensures that single point of contacts (SPOC) is assigned from each function, so as to provide faster and better support to the project.

Construction sites require different domain specialists to work in tandem like sales, designers, architects, products, quality, safety, commercial, expeditors, stores, compliances, etc. Each one deployed in the project has specific jobs, roles and purposes. There is a frequent tussle among team members in order to maintain proper balance among productivity, performance, quality and safety concurrently. This requires a great level of integration, understanding and mutual respect, keeping in view the bigger picture of the project. Any such internal gaps or misalignment among the team members could adversely affect the project.

3. Project Organisation Structure

A weak organisation structure causes delays, conflicts or failures as the key stakeholders may not be available as per the project requirements. Moreover, if the team members from different functions are not directly responsible or accountable to project, they may not take serious ownership and accountability of completing the assigned task on time, as different cross-functional team members support multiple projects, and thus they lack accountability or responsibility towards a particular project or project manager.

Many organisations follow the projectised structure, where all members from cross-functionals have dual reporting with primary reporting to the project manager and secondary reporting to their respective function heads. On the contrary, in case of matrix-reporting organisation structures, cross-functional team members have primary reporting to their respective function heads and secondary reporting to the project or construction manager. As a result, project activities and schedules are not given the required priority and attention. Cross-functional teams work as per their own priority, process, or functional SLAs (service level agreements). Even after the allocation of resources, their availability and accountability for projects, administrative approvals, performance reviews, etc. are decided by their respective functional managers, which becomes a major hurdle to meet the project timelines. Thus, the project organisation structure should be carefully decided, with adequate control and rights with the project management to assess the contribution of various cross-functional team members.

As a result, some of the risks experienced by the organisation structure relate to the following:-

Unclear accountability – CFT members may not have accountability, ownership, or priority of the project. They tend to prioritise their work accordingly to their convenience and willingness which impacts project performance.

Fear of resistance – Organisations do not provide open, transparent, and free environment for employees to contribute or discuss new ideas, thoughts, views, taking new risks and challenges. Thus they resist creativity and innovation.

Culture – Organisation culture does not promote recognition, respect, learning, knowledge, accountability, or growth for project team members.

Avoidance – Organisations avoid issues, problems or persons who communicate the problem, rather than understanding the problem and finding solutions; they side-line the person and thereby setting such examples for other employees.

Thus, it is recommended that the entity should duly evaluate the pros and cons of organisation structures, wherein all team members (allocated to projects) should have priority, accountability, and ownership for the project. Furthermore, all backend or office team members from cross-functions must be exposed to customer meetings, real site conditions, issues, challenges and environment. This shall help to foster collaboration among them and exchange their perspective, priority, points of view and contribute more effectively.

4. Sites and Surrounding Risks

Usually, large construction sites are either green-fields or brownfields, which have a variety of risks, deterrents and challenges associated thereof, that start with ground-breaking at sites. Site land may have waste, barren, undulation, structures, crops, or many other movable or immovable encumbrances. Such field legacy concerns the construction team, as its removal and diversion involve a high amount of time, energy, costs and approvals from various stakeholders, agencies and departments. It may be difficult to precisely ascertain the efforts required to get rid of such hurdles.

Other key risks experienced by the teams at the site include:-
   a. **Land & access** – Acquisition and physical possession of encumbrances-free land is a major execution risk, which severely affects the projects. Even subsequent changes in the

land parcel, shifting of boundaries or layouts affect its techno-commercial feasibility, work continuity or lead to reworks.

Thus, organisations must ensure that lands are properly acquired, titles are transferred, boundaries are correctly earmarked as per correct records of land/revenue department and land utility converted. A robust mechanism is to be adopted to take physical possession of land from authentic owners to de-risk conflicts later. Knowing in advance about any ongoing disputes, litigations, compensation, exiting agricultural crops, structures, cart-ways or pathways, or any other demand, are to be sorted out before concluding land title transfers.

It is a common hurdle to receive hindrance-free site access to start the construction activities. Sometimes, despite acquiring lands from legitimate sources in documents, actual physical land may be encroached or hindered by other local people or political parties on the ground.

It is important to note that usually, project gestation and decision of order award takes a long time since the time of bidding, (especially in government jobs); then the land surveys are to be carried out again, before signing off the deal, if there is a long time gap between bidding and ordering. If sites, routes and approach surveys are not conducted thoroughly and risks involved are not envisaged correctly, it becomes cumbersome to commence and carry out field activities.

b. **Site hurdles** – The contractor should consider the time, resources and cost involved in removing or diversion of existing hurdles like trees or bushes, boulders, electric lines, culverts, tube-wells, huts, ponds, structures, crops and pumps, roads, cart-ways, water embankments (bandhs), pathways or crossings, and other local setups or operations. Many sites may also encounter existing underground electrical lines, water pipes, drains, gas-lines, telephones/internet lines, etc., which require proper planning to remove or divert. Sometimes, even government departments may not have proper drawings of underground cabling or telephones lines, which makes the situation worse for construction teams.

Certain sites also encounter religious or spiritual worship setups (temples, churches, mosque, graves, tombs, stones, idols, trees, etc.), which are difficult to shift. Rather these require proper access, barricading, or approach by local people. Such constraints should be well-considered in the design, layouts, costing and planning.

Removal or shifting of existing setups may require approvals and permissions from government agencies, which are time-consuming and involve costs and expenses. Therefore, the same must be agreed during contract negotiations and adequately provided in the costing.

c. **Customer approvals** – Contractors have a high dependency on the customer's approval for design, specifications, drawings, change requests, documents, and measurements. Based on customer approvals only, can all manufacturing and construction activities take place. The parties must agree on the procedures and SLA (service level agreements) for approval, rejection or revert on documents, as delay can impact performance and schedules. Thus, proper records and logbooks have to be maintained, so as to correctly ascertain the cause of delays, when needed for claims, time extension or waivers. Likewise, SLAs of internal cross-function responses should also be monitored by the construction team. In the event of an abnormal or frequent delay, escalation or red flags should be raised immediately to avoid consequences later.

d. **Material related** – Delayed or interrupted flow of material deliveries, shortages or defective material deliveries, etc. hampers the pace and performance of construction work at the site. These affect productivity (time loss) and cause unnecessary idle charges of labour and construction equipment. It is nostalgic to share the experiences that sometimes, even the delay in timely supply of minor items like foundation bolts or templates impacts heavily on constructions of base foundations of major equipment, towers, and other plants. Thus, close and regular monitoring of all material including hardware, components and accessories has to be carried out minutely so as to find the gaps in manufacturing, inspections, logistics, dispatches, custom clearances, unloading at the site, etc.

e. **Right-of-way** (ROW) – One of the major execution risks at construction sites and cannot be ignored. Right-of-way is a contentious issue and commonly disputed in India. Such issues arise when an individual or group assumes that the land leased or purchased, is either infringing within someone else's boundary or was public rather than private-owned land. Some land areas may also be claimed by more than a single owner and lead to disputes. There may be cases where landowners (mainly farmers or local people) re-assert their rights, even though it was legally acquired or sold by someone.

If the land is provided by the customer, the customer may be obliged to provide right-of-way to the contractor; else the contractor assumes all responsibility of resolving such issues. Ultimately, land free from hindrances, encumbrances and encroachments are required, so that execution work can be performed without interruption. Generally, such restricted or litigating sites have risks in acquiring or taking physical possession from farmers or local people in greenfield projects, which demand compensation, extra money, or demand employment. Due provisions related to the settlement of such payments, penalties, interests, and compensation have to be explicitly laid down in the contracts to de-risk any ambiguity or conflict.

f. **Performance or productivity risks** – Accordingly to the Labour Bureau, Ministry of Labour and Employment (GoI) statistics, there is a rapid increase and demand of workforce in India. Construction sites are highly dependent upon large numbers of local labour and workers. Availability, skill sets, competency and other local concerns significantly impact the planned productivity of projects. Local workforces are less predictable and committed, as they may not come to work regularly nor intimate in advance their absence plan. Moreover, even skilled workers do not possess adequate experience; thereby they need higher supervision or/and result in low productivity. Due to high work pressure and scarcity of time at construction sites, it is also not possible to give much training to these casual workers and/or retain them for long. Hence, their shortage and incompetency impact project productivity. Thus, the planners should envisage this turnaround factors and competencies in planning and consider contingency. During local festival seasons and social celebrations, workers tend to remain absent and it hampers workers' availability significantly. For example, a majority of the workers remain absent during festival seasons like Diwali, Durga Pooja, Holi, Navratri, and other local festivals; so the planner should not discount the low productivity during such phases.

g. **Local disturbances** – Local social, religious and political parties' interferences may also influence the pace of construction works and harm/affect the project performance. These parties or agencies may have various interests or demands related to employment for local people, vendors, service providers or demand donations. The risks with these local gangs are that they neither have adequate competencies nor are competitive in terms of cost and quality. However, construction companies are still compelled to negotiate and engage them at sites. Over a period of time, some local farmers and villagers have become very powerful in certain areas and they may pester or harass the contracting parties. So, it is important for the contractors to properly coordinate and liaison with such parties. Many contractors are forced by the local people to engage vehicles, equipment, construction material suppliers and other services from the local people. Besides, some local bodies may expect financial supports as well.

h. **Construction ability of sub-contractors** – Non-availability of good local sub-contractors, poor quality of workmanship and weak processes adherence of such vendors could be a risk to projects. Poor workmanship could lead to rework and wastage of time and cost. At times, there are compelling reasons or monopoly of the local conditions to engage these sub-contractors. Despite charging higher rates, they do not perform upto the required quality standards. Rather, such sub-contractors may create unnecessary extortion, threats and stress to the project teams. These are threats to the project timelines and budgets. Many sub-contractors shall not even comply with the statutory or compliance requirements.

i. **Synergy among sub-contractors** – Construction sites face a high degree of fragmentation as a large number of tasks, activities, resources, and teams work concurrently. A large number of these workforces requires a high level of coordination and cooperation. If not all, most of their tasks are inter-dependent on other agencies. For example, an electrician can work in a room, after civil masons have completed the structural work. Similarly, most electrical equipment erection can start after equipment foundation is constructed by the civil team. All these activities require detailed planning on each front, cohesiveness among the stakeholders and progression in line with the project schedule. This also creates a lot of conflicts and confusion among the teams and affects movement at the site. This becomes more challenging while many independent agencies are deployed.

Certain projects face cultural or language barriers also, as different teams are deployed from different parts of the geography at sites. Some projects may require international teams, where communication and interpretation become a hurdle. Despite diverse activities, a high amount of coherence is required at the project site to play the symphony well.

j. **Delayed payments** – Timely payment to suppliers and sub-contractors is an essence for the success of the project. Money is the bloodline of projects. However, contractors seldom get paid on time as per agreed payment terms. Delays in opening letters of credit, full payment on time or discrepancies in payments can have a devastating impact on the project performance. Non-receipt of payments by the contractors impacts their cash-flow and consequently, delays or defaults in payments to their down-line vendors, sub-contractors, workers, and employees. Non-opening of LC could procrastinate manufacturing or shipment of materials. To prevent or minimise the impact, parties may agree on the outcome of such delays including interest charges on delayed financial obligations, extension of time, suspension of work due to extended or abnormal defaults. The hardest measure of such delays could lead to termination, disputes, and litigations also.

k. **Permits & approvals** – Every project acquires certain licences and permits before starting or during the construction of the project work and even thereafter. Applicability of licences or permits may vary according to the place, activities, or applicable laws. Thus, the requirement of licences and permits should be agreed at the planning or contract-formation stage. For example, a construction project in power and renewable sector requires approval from the land and revenue department, labour licences, building (BOCW) permits, forest, DISCOMS, utilities, electrical inspectorate, Bureau of Indian Standards and other government departments. Many agencies can charge fees to qualify before issuing the licenses. However, due to scarcity of time in the project and too much time and follow-ups required in acquiring permits and licences, some contractors may ignore, forget, or miss the same, which may cause violations and warrant unnecessary compliance risks with penal actions.

l. **Field changes** – Despite the best designs, layouts and assumptions, projects still face changes due to multiple reasons including site soils, material constraints, and other conditions. Changes are frequent and an inevitable part of the construction projects and could be risks if not managed. These field or design changes can be triggered by customers, contractors, laws, other stakeholders, or conditions. Such changes could also be triggered due to errors, omissions, or conflicts within contracts and/or drawings or varied interpretations by parties. Most changes may affect project scope, timelines, or costs. A construction team must track all changes during the construction phase, establish change request procedure, and ensure that all changes are reflected in as-built drawings, contract amendments and handing-over documents. Many changes can cause workflow interruptions, disagreements, and ambiguity between parties. Thus, a systematic or prompt response mechanism may be established to sort out such issues, vagueness, or discrepancies, so that neither the project performance is affected, nor relationships spoiled among the parties or workforces.

m. **Thefts & pilferages** – Stealing, pilferages and thefts of material are not uncommon at construction sites. Such occurrences may also take place with the involvement of some workers or break-ins caused by insufficient security arrangements. In the case of distributed or scattered job requirements where centralised stores are unable to cater completely like power telecom cables, transmission lines, oil & gas pipelines jobs, the risks of loss increase drastically due to the widespread areas and inadequately protected storage yards. Thus, the construction teams must ensure that proper storage yards, storerooms, boundaries, lock and keys and security arrangements are made, including CCTV or manual watch and ward with adequate protection. The risk register should reflect the same appropriately.

n. **External hazards** – Extreme weather or atmospheric conditions such as excess rains, cold, snow, fog, heat, storms, floods, earthquakes, etc. can have a direct impact on the normal site execution activities. In addition, the other external hazards activities like vandalism, sabotage, interferences by local influencers, labour strikes, terrorism, social commotion or unrest type of risks can affect the project. Exceptionally, even local public administration or local police may not be of much help due to excess local people's pressure or political reasons.

o. **Back-to-back basis terms** – Most contractors receive the order from intermediate customers, who are not the actual owner, end-users, or investors themselves. These contracts are generally bound to honour the terms agreed by a customer with its end-user or owners, where the terms, risks and responsibilities enumerated under main contracts can be different or higher. So, contractors are required to satisfy the customer and the end-user or owners, which can be subjective and complex.

Further, for example most renewables projects are undertaken by developers based on their power purchase agreements (PPAs) with government agencies. These PPAs enforce many obligations on the developers prior, during and after the construction of the plant. Developers generally download most of their obligations, responsibilities and risks to their contractors. These obligations may require extensive liaisoning, time and costs. In case the contractor fails to comply with those obligations, the contractor is liable to compensate such costs and penalties to the customers. So, proper envision and consideration have to be made by the contractor in its costing and execution planning.

5. Communication Plans

Disintegration among project stakeholders is a high risk for projects. If some stakeholders think that you are building a switchyard but you are actually constructing a solar power plant, then certainly there is a big communication gap. Likewise, if some stakeholders think that you are constructing a concrete room, while you actually plan to use pre-fab structures or porta-cabins, then this is a communication gap. Such projects are bound to fail. First and foremost, all key project stakeholders must be on the common platform of scope, deliveries, challenges and risks.

Projects are all about communication, information, data and follow-up management. In the absence of proper communication plan or communication matrix, there is a high probability of miscommunications, misinterpretations or misunderstandings of information and data. Therefore, clear communication matrix should be agreed between the parties from the beginning, highlighting roles, names, titles, contact details and escalation matrix of all stakeholders in the project. Roles, responsibility and authorities of each team member should also be agreed and defined in the project, so as to have a fair and smooth flow of communication and avoid any miscommunication, or misconception at a later stage.

Communication plans should also define the cadence and frequency of various meetings and reviews required at different locations and platforms.

Although communication is crucial, relevant intimation or data should be shared with the relevant stakeholders. In the present digital world, there is a tendency to broadcast (reply all) emails to everyone, whether it is required or not. So communication etiquette or guidelines must be defined as per the purpose and expectations from stakeholders. Contracting parties can also decide the RACI matrix for the normal exchange of communication. Also remember that confidentiality and protection of intellectual property right are important and the responsibility of all team members; so too liberal and casual flow of communication should be controlled in projects.

6. Monitoring & Controls

There is a popular saying that "*anything that can be monitored can be improved.*" Customers, investors, stakeholders, and leaders regularly need project status or updates. To meet such requirement, project progress should be monitored from the beginning and throughout the project lifecycle. If the actual time, resources and efforts spent on each task, activity or milestones are not tracked properly and compared with plans (or estimations), it will be difficult to present the correct picture to the management, which can lead to incorrect inferences and decision-making. One can provide the correct update to management if the same is monitored regularly and reflected as to what is completed and what is pending. This will help to re-plan the balance resources and times for completion of the balance works. In case, some actions require acceleration or a different strategy, the same can be worked out suitably before it is too late.

Project Management Office (PMO) should decide on enabling project organisations to embed the best industry practices, leveraging project lessons (besides capturing and compiling databases), and facilitating projects and leaders where projects can flourish. They should standardise plans, processes, reports, cadence, and review mechanism for projects, by anticipating the realistic conditions. This also includes the formats and frequency of reports and dashboards, which shall be shared with stakeholders. This helps to apprise stakeholders about the progress and gaps so as to seek their support. The team should also plan reviews with cross-functional teams, customers, contractors, or other agencies involved or impact the project. During the reviews, teams can come out with many new ideas, risks, and actions, which can help to take suitably corrective or prevention actions. Documentation, recording, circulation, archiving and retrieval of data are also an important part of central PMO responsibilities.

7. Return on Experience (Lessons Learnt)

Every engineering and construction site is unique, even if it replicates the same design, layouts, drawings, specifications, standards, processes, and teams, as the variable factors and site conditions are different. There will always be a lot of stories to share and tales to tell at every project site. These entail a lot of issues, challenges, failures, situations, problems, success, and solutions throughout the project life cycle. There can be mixed learning – the positive and the negative. Learning is a continuous process of timely capturing, documenting, analysing, storing, and ensuring retrieval of lessons learnt.

In the current intensely competitive time, no organisation can afford to repeat the same mistake again at all projects. Thus, lessons from each project are to be captured and shared with other project sites. It can bring tremendous benefit for the organisation and save a lot of cost of reworks and poor performances. This can help to stimulate productivity by using proven methods and avoid incorrect practices.

Each project gives a lot of lessons and experiences to the team and organisation. It is the downside of the project, if the organisation fails to methodically identify, capture, share with concern team (function) and leverage the lessons of each and every project. It requires proper mechanisms or platforms or meetings to capture the learning or carry out the post-mortem of the key learnings.

The complete project team—stakeholders, cross-functional members and parties can contribute to identify the lessons and exchange with the teams. The team should take note of what went well and what did not, what worked out to be a catalyst for outstanding outcomes and what was missed or misunderstood by teams.

Be mindful that the objective of such exercises is never to fault-find, defame, blame, implicate or cause conflict between individuals, teams, or function, but to learn and imbibe the best practices and lessons for future projects. These inputs, lessons and experiences guide to make the estimates more precise and realistic for future projects.

To mitigate high risks and deterrents in the construction industry, companies are harnessing new technological innovations and the latest trends of doing the work cheaper, faster, better and safer. Some of such changes or developments include:

a. **Integrated contracts** – Most customers, developers and investors prefer to execute projects with turnkey contracting model with a single party, thereby entrusting the full risk, responsibility, and accountability. This saves a lot of time and hassle for customers to interact with multiple contractors, repetitive documents, approvals, resolve their conflicts and disputes regularly and monitor the progress and gaps closely.

b. **Modular structures** – With rapidly growing demands and shrinking timelines in construction, many architects, designers and construction specialists are banking upon pre-cast or pre-fabricated structures of walls, roofs, beams and pillars. This reduces the site risks and increases the quality, safety, productivity and performances.

c. **Drones** – In large project sites, drones are an easy, fast, and accurate system to track the real-time aerial views, heat-maps and thermal images of material and site fronts. Drone survey outcomes or data drive actions help for rapid decision-making and streamlining process gaps, wherever encountered by teams. Moreover, this gives the flexibility to monitor data, reports, images, and video from a wide geography regardless of physical presence at the site. Drones can also be used to bridge the security gaps, minimise human biases or inventions by keeping a continuous vigilance over large land parcels.

d. **Smartphones & cloud meetings** – These inventions are proving a boon for sites. With smart mobiles and videoconferences, it is far easier and accurate to track construction statuses, establish accountability and monitor the progress on a minute basis. Many companies are exploiting the technologies to capture the truth of the moment by seeking momentary or daily progress reports with pictures and video clips. Further, the use of conference tools like Zoom, Teams, Skype have become an integral part of the meetings and experts can extend support from anywhere globally.

e. **High-end equipment** – The construction sector is witnessing great improvement in the use of construction equipment and tools, which are better, faster, safer, and leaner. This equipment also saves time, noise, vibration and damages from pollution while helping to reduce construction site accidents and saving on poor quality works. These are not only reducing the risks of human errors but benefitting in an improved environment, quality work and speed of productivity.

Concluding, despite construction sites posing inherently the highest risks, this industry is still booming and growing rapidly. The main reasons are the high demand, considerable expansion globally and continuous technological developments in this sector. This fact can also not be neglected or denied that this constructions segment is facing an acute pressure of rising construction costs, enormous labour issues, local disruptions, complex regulations and shrinking margins. Thus, the cost of efforts employed must yield the desired outcomes.

Chapter 9

HEALTH, SAFETY & ENVIRONMENT RISKS

"Safety overrules Quality, Schedule and Budget"...

Universally and legally, everyone has the right to life and liberty. The fundamental need in any business or project is that *"Everyone in the workforce must get back home safely after work."* However, the reality of the construction world is far away to achieve this basic need in its entirety, as completion of the work gets the highest priority and emphasis in construction sites thereby leading to compromises of human health and safety. Albeit the construction sites operate in discrete, diverse and complex conditions as compared to manufacturing and other sectors. These site conditions contribute to the majority of construction accidents. Ironically, despite being part of many organisations' vision and policies, HSE issues do not receive the required attention and focus as projected or warranted. Indeed, the fatalities in developed countries are less than half as compared to construction conditions in India.

Deployment of a high number of resources and gangs concurrently by multiple agencies require a high level of coordination and cooperation. Though collectively it is called a project but actually it has a large number of sub-projects and activities embedded within. As a result, this poses the highest risks and may significantly impact human health, safety and environment under which workforces operate. Most accidents and casualties at construction sites are attributed to unsafe acts, poor practices, malfunctioning, negligence, or ignorance. It may be considered as sub-standard working conditions.

Normally, working conditions have many threats and deterrents due to tough work requirements and potentially dangerous situations. These situations demand high physical workloads, mental pressures and stretched work timings at construction sites, thereby increasing the risk of their health and safety. Not only the hazardous surroundings at sites, but also workers are exposed to extreme and inclement weather conditions (excess heat, cold, winds, rains, sand dunes, snow, etc.), natural disasters, pollutions, and other similar issues. Nonetheless, some obscure sites may also have the risk of wild animals, underground insects, reptiles, etc., which are serious threats for workforces, which not only require immediate first-aid but specialised treatment also.

The construction sector provides employment to a large number of workers. In India, almost 49 million people, who contribute to over 7.5% of the national gross domestic production (GDP) are employed in the construction sector. However, this is still one of the most neglected and perilous sectors for workforces. Despite playing a vital role in the country's economy, the workforce in this sector is unprotected as compared to many other sectors. It is alarming to note that some statistics reveal that the possibility of fatalities in the construction sector is much higher than the manufacturing industry and the risk of major injuries is almost double. Thus, it is the prime responsibility of all employers to seriously consider implementing the robust HSE system at construction sites, as part of their social and sustainable responsibility. Rather continuous evolution, awareness, trainings and enforcement of compliances and supervisions should not be compromised at all.

International Labour Organisation (ILO) statistics of 2017 revealed that 48,000 people died in India due to occupational accidents and the construction sector contributed to almost 24% of the fatalities. Besides fatalities, a large number of workers suffer from health hazards and non-fatal injuries. These statistics and severity are mind-boggling to visualise the risk involved in construction sites and the need for safety and health of the workforces.

Construction sites are precarious and dynamic, where perils and exposures continue to increase, change, and spread across the project premises. Besides the workforce, construction sites also utilise a large number of heavy vehicles, tools and machines for various activities. A large number of moving vehicles and machines are used for lifting, and shifting the equipment and material (like cranes, forklifts, boom lifts, hydra, hoists, tractors), excavators and diggers (like DTH, ajax, bulldozers, JCBs, backhoe, crawlers, concrete mixers), electrical and power tools (like MLD, DG sets, welding machines, pumps, motors, gas cutters). These are high-risk activities and when two or more machines or activities are implemented simultaneously, the risk exposure increases manifold. For example, painting of beams or long pillars. This includes high-risk activities of working at heights, use of inflammable paints and fall of man and material. This increases the risk manifold for the people working underneath.

Further, most construction lands are uneven surfaces, ambiguous and movement of vehicles (carrying equipment, material, or workers) have a higher probability of accidents and casualties for people and properties.

As a result of the above, constructions workers run the risk of not only lower work ability but also lower health conditions. Practically, long-term construction workers commonly suffer from multiple health problems like physical (musculoskeletal, fractures, radiation, hearing), psychological (like distress, depression, mental disorder), respiratory diseases (like asthma, pneumonia, pulmonary, cancer), Skin diseases (like blisters, itching, cracks, swelling, thickness), noise and vibrations (like Sleep disturbance, hearing loss, blood pressures, nerves and blood vessels disorders) and many others.

In spite of high risks and poor working conditions, every accident and injury is preventable by adopting good safety practices, loss prevention approach, systematic evaluation and implementation of risks at sites.

Some reasons contributing to high HSE risks are:-

- Lack of commitment from the leadership
- Highest priority to the speed of the work, rather than safety or environment
- Tough terrain, severe weather and remote locations
- A high number of uneducated or under-educated local workers
- Shortage of qualified and competent HSE motivators and professionals at sites
- Inadequate sources of knowledge and learning of site teams about changes and development taking globally in the safety domain
- Lack of adequate facilities to meet HSE legal requirements and
- Missing signages, posters and safety symbols

Each organisation, including the leadership, must commit itself to prevent these risks to human life and properties at every site. The organisation leaders must devise a comprehensive safety management system to ensure the safety of all employees, workers, personnel and visitors at sites. Here, atomisation of the construction industry is the key, where we shall be able to reduce the exposure, if not the hazard.

The organisation must prepare and implement a robust HSE Management System, with support and commitment from the leadership. The HSE system should cover the core objectives:-

- Commitment to **"Zero Injury"** or **"Zero Accident"**
- Implement the HSE policy with accountability
- Define HSE Organisation, responsibilities and resources allocation
- Streamline Risk Identification, Evaluation and Management processes
- Availability and adherence to **"appropriate 100% PPEs"** at Sites
- Train and sensitise the workforce to reduce incidents
- Enforce compliances to legal and statutory requirements related to Labour Laws, Environment and Organisation HSE policy
- Correct and timely incident recording
- Continuous implementation of HSE Policies, Supervision and Monitoring
- Audits and Reviews
- To implement the above objectives, contractors should devise processes for reporting Near-Miss or Incident without implicating or intimidating individuals. In fact, hiding the facts or incidents should be considered as a violation of "Integrity." Good implementation and management of such processes should be appreciated, and such lessons are recorded and used while refining the processes and policies.

The team must remember that:-

- Every incident is an opportunity of learning and not to find fault with individuals
- Establish a mechanism to identify, validate and document the monitoring and reporting incidents in a professional manner
- Positive behaviour – no blame-game, no-victimisation, or no-punishment policy
- Platforms to share the lessons learnt, give feedback and bridge gaps
- Promote or recognise the HSE motivators to enhance acceptability across the teams

Implementation of the HSE policy requires a high commitment and passion in any organisation, which can be driven by making the HSE a part of their primary deliverables or key performance indicators. This can facilitate to inculcate a culture of collective or collaborative implementation for a wider goal and objectives.

The HSE team along with other site team members should frequently visit and assess the working area risks even before the resources (manpower or material) reach or the work is commenced. This process shall support to identify the locational risks. There may be high-risk situations or locations, where the work area is used for trespassing, grazing animals, or even crossing high-tension electric lines, which are low enough to electrocute the excavators or workers.

While implementing the HSE policy, the team should carefully assess:-

- Site risks and operational feasibility before the start of works
- How safely people and material are going to reach (route surveys), move around (site surveys) and the surroundings areas (security situations)

- Site fronts, places, substances, or local situations which can be perilous

- Routine and non-routine activities or operations to be carried out

- Evaluate the risks of common diseases (caused by local conditions, water or pollutions), epidemics, infections, etc. in near-vicinity or environment.

Construction leaders must comply with the applicable labour laws and prohibit the deployment of children, unauthorised migrants, pregnant or expectant female workers and people with any disability.

Once the hazards are identified, the HSE experts should advise or guide as to how these can be best avoided or mitigated by the workers, so as to reduce the probability of accidents or causalities. The construction teams should maintain proper records and registers of all incidents (near-miss events, fatal and non-fatal) and apprise the leadership.

At construction sites, even the most effective controls may not work in the absence of strict supervision or disciplines. So merely providing PPEs, documenting processes and imparting training may fail to meet the objectives. This will require proper supervision, reviews and audits of the work, as a regular part of the process implementation. This must be made a part of the habits and culture of the organisation.

Before Start of Work

An EHS specialist should perform the following activities before the start of new construction sites:-

- Visit complete site, surroundings and all fronts, identify and access the hazards and risks

- Devise EHS Plans, Strategy, Registers, Checklists and Formats

- Ensure the availability of PPEs
- Prepare Training Calendars—Site and job-specific activities
- Organise First-aid kits
- Establish controlled access system/attendance/permit to work
- Finalise and fix safety signboards, posters, symbols at main and crucial points
- Identify local emergency contact details (Police, Fire, Hospital, etc.)
- Prepare safety-work procedures, methods and implement the plan

After Start of Work

Upon the arrival of the project team members, contractors, and visitors, the EHS specialist should regularly perform the following activities at the construction sites:-

- Induction training to all workers and visitors (covering site layouts, prohibited and permitted areas, main hazards and key precautions during work or visit)
- Inspect the tidiness of cluttered material and segregation of good and waste material – including storage processes related to loading, unloading, storages and material movement, within the sites through proper routes and with supervision
- Inspect work fronts (height works, excavation, foundations, holes and trenches, etc.)
- Check conditions and calibrations of construction machines and material (jacks, pulleys, scaffoldings, electrical wirings and arrangements, power tools, etc.)
- Hazard hunts with teams to detect any impediments or hindrances or potential for risks or restrictions of work

- Ensure availability and training for use of correct PPEs (all sizes and types)
- Ensure safe dismantling procedures and segregation of dismantled stuff
- Manage and monitor moving vehicles and objects (verify vehicle and driver's documents)
- Demarcations/barricading of risky zones or dangerous areas
- Establish permit to work (PTW) process

Several thousand workers meet with accidents or get injured every year in the construction industry. The root cause of almost 90% of such accidents or casualties can be attributed to unsafe acts, negligence or ignorance by teams or workers.

Most common risks and hazards countered at construction sites include:

a. **Working at heights** – Construction sites involve risk of working at heights. As the buildings, structures or towers go up, the risk of falling increases. These are the major causes of accidents and casualties. The risk includes working from roofs, scaffolds, ladders, cradles, tower structures, etc. Contributory factors include negligence, weak scaffolding, equipment failure, restricted access, or mobility, etc.

Prevention – Correct use of PPE (helmets, harnesses, gloves), provide guardrails, edge protection, secure ladders or scaffolding, tidy tools and materials on roofs and elevated walkways and avoid working in nights and extreme weather conditions.

b. **Moving objects** – Sites are extremely busy, hazy with the bustling of activities. Multiple vehicles, equipment and machines keep moving concurrently.

Prevention – Hazard zones and routes for vehicle movements to be clearly marked, minimise people's movement on the way or close to moving vehicles, adequate use of warning horns, lights, sirens, (with backlights and back horns), use of PPEs, especially reflectors, safety coats, glowing jackets, etc. as these are seen from distances. Ensure that drivers with valid licences operate the vehicles and not helpers or workers.

c. **Slips, trips and falls** – Most common hazards at sites. Project sites are uneven surfaces during the construction phase. All across the sites, there may be a lot of unused heaps of material or wastes, open trenches/holes, waterlogs, and wet patches. So, the risk of slipping, tripping, and falling increases.

Prevention – use of correct PPEs, regular attendance of workers (entry and exit timings), make clean and clear pathways for the movement of people, keeping material in a tidy and non-scattered manner in stores or barricaded areas, spread of wastes and scraps to be removed immediately and kept aside, remove logged water, muddy and slippery surfaces be treated or sign-post put clearly.

d. **Noise hazards** – Common problem at most sites, though the noise levels can vary at different locations. This is a commonly ignored hazard and protective equipment provided to workers are either of inadequate levels or avoided from the use by workers. Loud, repetitive, and excessive noise causes long-term hearing loss or deafness to workers. Noise can be a dangerous distraction too for all workers and visitors.

Prevention – appropriate PPEs to be provided to workers.

e. **Airborne hazards** – Airborne dust, fumes and pollution are a silent killer. Sites are exposed to environmental hazards (dust, storms, and sand dunes) and itself produce dust, pollution and wastes due to stone masonry work, demolition of old structures, removal of rubble, woods and tiles dust, and spread of civil works material, etc. This dust may be invisible, fine and toxic mixture of hazardous material and fibres. This can directly damage workers' lungs and lead to chronic diseases related to breathing and heart-related problems.

Prevention – use of face masks, helmets, suppress dust with water, tarpaulin or work in covered areas.

f. **Material handling** – manual handling, lifting, holding, moving, pushing, pulling, or carrying the material by workers from one point to another is an unavoidable requirement of all sites. Handling of material carries a high degree of risks for workers like bruises, fractures, wounds, laceration, damages to muscles, tissues, discs, ligaments, nerves, etc. The intensity of injury due to material handling could be mild-to-fatal.

Prevention – use of correct PPEs, adequate training, and limitation for manual handling of heavy material, regular supervision and monitoring of workers. Avoid movement at uneven, wet, muddy or dark surfaces.

g. **Collapsing** – As the buildings, towers or equipment structures go up, the risk of falling things increases. This causes the risks of destroying existing structures, setups, material, or building. There are high chances of workers falling in open excavations, trenches, sumps, holes and getting injured. Even the risk of collapsing of unstable heaps of earth (or muds) along the trenches on the workers inside can cause harm. Many power plants have a high number of open trenches (for laying of cables and pipes) and foundations (for building and equipment) which could endanger human life and moving vehicles.

Preventions – Proper delineation, barricading and securing risky areas, deploy proper retaining barriers and scaffolding, inspection and sufficient lighting arrangement during night works.

h. **Electrical hazards** – Sites use a large number of tools, tackles and machines that require a temporary power supply to operate. Usually, electrically charged cables, wires and extension cords are spotted and spread all across, most in shattered conditions. Though these are temporary, they are used for long periods leading to deterioration in their conditions and quality as these are subject to traffic abuse, vehicles cross over them, which can wear down their insulation and create shock hazards. Such electrical wires and cables are not only stripped off but also overdraw power from circuits, causing heat-up and fire. These are risks for not only the electricians or technical staff but also for all the workers and visitors at the site.

Prevention – Proper risk assessment of such cables, wires, extension cords and tools deployed, use by qualified electricians or under their supervision, necessary use of PPE (gloves, safety shoes, safety goggles and protections like LOTO, etc.).

i. **Harmful materials** – Many sites require demolition of old existing buildings, sheds, or structures as part of the construction. Such sites expose the workers to the risk of asbestos, hazardous waste, scraps and chemicals (adhesives, solvents, liquids, paints, thinners, varnishes, glues, etc.). Handling of such material can be harmful to the workers and may cause severe diseases of lungs, mouths, eyes, skins, pulmonary system and others.

Prevention – Use correct PPEs, regular health check-ups, proper place and procedures for quick removal and disposal of waste and scrap, good practice protocols in the event of leaks, spills, spread and accidents of harmful materials.

In order to prevent accidents and causalities, the construction team must ensure that the following main precautions and preventions are adhered to:-

a. **Use Personal Protective Equipment (PPE)** – This is the last line of defence for all workers at construction sites. It must be made mandatory to wear correct safety gears and gadgets. Before the use of PPE, the employer is to ensure that the provided PPEs are of good quality, adequate in numbers and sizes. Ironically, many organisations make tall claims about compliances, but the ground reality is different. They compromise on the cost and quality of the PPEs and supply inferior gears or small quantities, which do not fulfil the requirements. The common excuse is that the PPE cost was missed in the budget. So, regardless of the reasons or excuses, adequate PPEs must be provided from the first day of commencement of work at sites. The employer must ensure that all workers are provided with correct sizes and a variety of PPEs and they use it during work. Many workers do not feel or find it comfortable wearing the PPE while working and the prime reasons can be their mindset, habit or behaviour. Thus, it is the collective responsibility of the construction and HSE teams to continuously educate, train and sensitise the workers about the benefits of using the PPE and how they can prevent accidents, injuries, illness, casualties and protect themselves. Eventually, the use of PPE cannot be left to the comfort or discretion of workers.

b. **Safety signs** – Practically, it is impossible to remove all risks and hazards at any project sites. Thus, different types of safety signages are marked or placed at various places or locations to communicate and raise alerts while people enter or work at sites. Certain signages or symbols denote specific hazards or potentially risky areas, where unauthorised persons should not enter, or authorised persons should also enter or work with specific or additional protections or precautions. Globally, there are thousands of safety signs, however, construction site-related safety signs are grouped in five categories – (i) Prohibition Signs, (ii) Mandatory Signs, (iii) Warning Signs, (iv) Safe Work Conditions, (v) Fire Safety Signs. These signs denote caution against potential hazards at a work front, area, or route. That may require special protection, care or additional precautions while planning to work. It may be the legal and statutory requirement that all constructions sites must put the relevant safety signs at the site through banners, boards, posters, etc. and spread awareness. Safety signs should be uniform, clear and concise. However, unclear or damaged safety signs could be misleading or confusing and may become funny reading material or a point of jokes among workers. Further, if possible, arrangements must be made to translate safety signs in local or regional languages so that the workers and labour could easily understand and adhere to.

c. **Safety culture** – It is true that values, beliefs, and habits that are practised by leaders (about health and safety) are actually driven to the lowest level at construction sites. It is the outcome of individual and organisation core values, attitudes, behaviours and perception that ascertain their commitments towards health, safety and environment. Culture is routed much deeper, like an iceberg, than what is actually visible above the water. Thus, it must roll down from top leadership. Many workers have a mentality of "monkey see, monkey do." If some leaders, managers, or supervisors do any unsafe act or negligence, the workers tend to follow or refer to the same too. Therefore, leaders set the tone and culture by demonstrating their commitments and respect to the use of safety gadgets (PPE). This proves the seriousness, attention and care for the team by leaders. Great organisations start their meetings, reviews, and discussions by addressing the issues of safety compliances, gaps and actions taken or required.

d. **Safety ambassadors** – Safety is not the responsibility of safety officers only but is the collective responsibility of all employees working at the site, be it a manager, engineer, supervisor, technician, or accountant and others. Every employee and worker of the organisation is a safety ambassador. Everyone at the site can demonstrate the authority to stop-work in case of violation of safety rules. No single person can implement safety at site or organisation. Come what may, each project site must have safety mentors or safety leaders, who are responsible to develop safety systems, procedures, culture and support in implementing the same in a safe manner.

With the recent outbreak of pandemic Covid-19, additional safety measures that have evolved as part of society's culture and at construction sites to avoid the spread of contamination among workers are:

a. **Hygiene & cleanliness** – Special focus and attention are given to the hygiene and cleanliness of sites with proper sanitisation of workplaces, cleaning, and hand-wash of individuals.

b. **Physical distancing** – The workers and staff should maintain a certain distance and avoid physical contacts, to the extent possible to control spread of contamination or infection.

c. **Face covers** – Using face mask to cover your mouth and nose, so as to avoid the spread of infections and virus among other people.

d. **Temperature scanning** – Body temperature of all workers, staff and visitors to be checked at entry/exit points of the project premises to control the spread of any infection from the infected person.

e. **Restricted movements** – To the extent possible, restrict unnecessary movements of workers, staffs, visitors, guests, and other non-essential team members.

f. **Segregation** – Maintain isolation or segregation places for workers, in case of any exceptional eventuality of suspect cases to be encountered at sites. Also have adequate provisions for first-aid medicines and emergency contact details of nearby doctors and hospitals.

Environmental Risks:

Environmental risks are the most unpredictable, unknown, and volatile risks that may affect the project performance and can also impact the environment as the outcome of project performance. This is a universal truth and can neither be hidden nor ignored. Environmental risks do not care about local, national or global boundaries. These risks may emanate from various external factors and can jeopardise the construction or operation. Eventually, if not managed well, environmental risks could be most frightening for society at large.

Normally some contractors assume the inclement weather or *force majeure* occurrences as environmental risks, whereas, the scope, scale and domain of environmental risks are much greater, which can become a serious threat for projects and society. Thus, detailed risk identification, analysis and mitigation are critical from the start of the project, until completion and operations phase of the projects.

Environmental risks have been increasing rapidly in speed, scale and scope. The tragic consequences caused to health and life due to environmental hazards are mostly irreversible. These risks are not only an irreparable aberration but also turning as formidable challenges for the stakeholders to find different ways and means to minimise or mitigate the impacts thereof. At the same time, these risks are causing additional commercial burdens and environmental risk liabilities on stakeholders in restoration and re-habitation.

With increased scope of environmental regulations and strict liability imposed by various government and statutory authorities, even organisations can be held liable if they cause damage to the environment, cause pollutions or endanger the life of living beings (human, animals, birds, reptile, plants etc.). The enhanced public awareness about environment and adverse consequences has escalated disputes and claims due to environmental damage in the construction sector. The environmental risks in projects depend on the site locations and surroundings, which can cover surrounding lands, localities, and proximity to protect species or natural habitats.

Construction sites are generally not environment-friendly and one of the main contributors to environmental disruption and pollution. Construction hazards contribute to loss of natural assets, severe stresses, and impacts the environment. In case the construction activities are not managed properly, it ends up damaging the natural and environmental heritage like rivers, lakes, ponds, aquatic life, wildlife habitats, and other species. This can disrupt wildlife habitats, spread contamination, and harm ecosystems and environmental balance.

In fact, this is one of the biggest responsibilities of the associated stakeholders to achieve complete sustainability and protect the environment for future generations. Regardless of the role of any individual or functions in projects, it is the collective social obligation of all to achieve this objective for higher and bigger benefits. This requires a significant contribution from all stakeholders including lenders, investors, developers, architects, consultants, contractors, manufacturers, HSE experts, and concerned government bodies to work in a cohesive and cooperative manner to overcome this formidable challenge.

There is a continuous need to strike a balance between rapid development and growing environmental concerns. Together with HSE specialists, the construction team can envision the risks involved and find out ways and means to curb the same. Some of the environmental risks involved are:

a. **Soil degradation** – Most green-field and brown-field projects necessitate extensive land disturbances, which may involve the removal of agriculture, vegetation, plantation, and reshaping of topography. These activities may cause the soil to be vulnerable to erosion. Earth cuttings, shifting, filling, removal or excavation which may cause many problems to earth, air, water, dust, and affect the environment with long-lasting impacts. As a result, soil erosion and loose earth can also be detrimental to the groundwater and surrounding waterbodies due to siltation. This can result in mud floods and flash floods during heavy downpours.

b. **Disruption to ecosystem** – Sites can cause damages to various species of animals, birds and plant life, and other habitats. Strong ecology assessment, risk analysis and mitigation action plans are required to ensure minimal losses during the construction as well as operations phase. It may require many actions including change of project sites, change of methodologies, or diverting routes or replenishments, where feasible. This is more crucial especially if the construction is planned in the close vicinity of any animal sanctuary, forest, or catchment area. To the extent possible, ecological losses must be reduced and suitable protection taken for the conservation of the flora and fauna.

c. **Water pollution** – Availability of groundwater for construction is a big risk and challenge at construction sites. This is a formidable challenge in the desert and dry areas. Also, the groundwater gets polluted with the use of various types of wastes and materials at sites like paints, glues, adhesives, oils, diesel, chemicals, cement, etc. This contaminated water reaches the nearby waterbodies, ponds, drains and reservoirs, which is then used or consumed for domestic purpose, and by animals and pets. As such, water-borne diseases are common and high in India. Water quality is important for health, economic, ecological, and aesthetic purposes. Deterioration in water quality may affect the aesthetic values and may prevent uses of the water in regular courses. At construction sites, the removal of vegetation, area grading and removal of soils can cause great risk to water quality as these areas become susceptible to erosion. During the rainy season, the impacts and risks increase manifolds. Many sites generate a lot of wastewater, which is not disposed of properly or disturbs the natural follow of *nallahs* and waterways.

d. **Air pollution** – Airborne contaminants and dust particles spread (carried by the wind) all around the construction site and surrounding neighbourhoods affecting the air quality. Such pollutants or toxins spread in the air in a short time. The air pollution at construction sites is mainly caused due to burning of vegetation and wastes, emission of fumes, smoke and gases due to the excess use of diesel vehicles or machines, the release of chemical impurities such as heavy metals, acids and other toxic bases. Air quality gets impacted with fragmented dust particles in the atmosphere caused by demolition, cutting, grading, filling, removals, fabrications, and other construction activities. Air pollution is also caused by emissions from the movement of a large number of heavy vehicles at constructions sites.

e. **Noise & vibration** – It is caused due to numerous activities like the use of heavy construction equipment, tools, machines and vehicles' movement. Noise and vibration levels vary at different constructions locations depending on the types of equipment used and operating mode. Adverse impacts resulting from construction noise and vibration are generally limited to areas adjacent to the project site and more specifically impact the workers and labour engaged at sites.

In addition, some constructions sites have hazards which are flammable, radioactive, toxic, corrosive or even explosive too. So, proper plans and mitigations are required to handle such high risks.

Although construction sites are full of environmental risks, adherence to good practices can contribute to control or minimise the damages. It requires thorough identification and assessment of the project sites, surroundings and likely construction activities which can impact the environment. Some of the measures, which can be implemented at construction sites by stakeholders include:

No.	Protection Measures	Control Benefits
1.	Restrict the burning of packing material, scrap, wastes or vegetation	Air Pollution
2.	Use new technology equipment for digging, excavations or operating machines to reduce diesel consumption	Air Pollution
3.	Reduce the spreading of dust at the site by using water sprays or sprinklers, tarpaulins on civil material and concrete works	Air Pollution
4.	Cover or protect areas which are generating dust	Air Pollution
5.	Fully cover trucks, trollies, tractors and other vehicles while carrying civil construction materials like soil, sand, powders, silts, gravels, bricks, etc.	Air Pollution

No.	Protection Measures	Control Benefits
6.	Do not allow littering or cluttering of raw material at the site—soil, sand, gravel, cement, chemicals, etc. and plan to reduce the time gap between material receipt and consumption.	Air Pollution
7.	Explore material sourcing close to the site to minimise diesel consumptions	Air Pollution
8.	Avoid unnecessary storage, spreading and waterlogging at construction sites	Water Pollution
9.	Wastewater disposal system must be planned properly	Water Pollution
10.	Cover trenches, foundations, drains and low areas to avoid waterlogging and damages of material	Water Pollution
11.	Minimise use of water For Solar modules' cleaning and use robot cleaning systems to save wastage of gallons of water	Water Pollution
12.	Use modern hybrid technology tools, machines and equipment to minimise noises, vibrations and also reduce diesel usage	Noise Pollution
13.	Explore noise-proof barriers (if possible) to restrict the spread of noise in limited working areas, affecting limited workers also	Noise Pollution
14.	Timely preventive and corrective maintenance of tools, machines and equipment so that major risks to life and environment are mitigated	Noise Pollution
15.	Stop or disconnect tools, machines and equipment, when not in use	Noise Pollution
16.	Use non-toxic paints, solvents, adhesives, glues and other chemicals	Chemical Pollution
17.	Properly dispose off masks, packing material, wastes, scraps, equipment oils, chemicals and other material (as per Government directives)	Chemical Pollution

No.	Protection Measures	Control Benefits
18	Segregate or properly store/cover inflammable or toxic substances like diesel, oils, paints and chemicals	Chemical Pollution
19	Minimise earth cutting, soil disturbances or unnecessary vegetation cuttings for mere aesthetics or beautification	Soil Disturbance
20	Demolition of existing structures or building or infrastructure	Soil Disturbance

Chapter 10

LEGAL & COMPLIANCES RISKS

"Until the contract is signed, nothing is real."

– By Glenn Danzig

Legal risks may emerge from contracts, conduct of legal and regulatory obligations and errors or omissions in fulfilling statutory compliance requirements. There are higher probabilities that the organisations may fail to comply with, deviate, or violate any of the applicable legal, regulatory or compliance requirements, which may disrupt the performance or increase the liabilities manifolds and lead to legal risks—like non-compliance, ambiguities, frustration, impossibility, conflicts, disputes and/or litigations.

All games require that its players should know the "***rules of the game***," to enhance the probability of success whereas, in contracts and laws, players should know well the "***game of the rules***." Generally, the constructions teams are not very legal or compliance-savvy; however, they need to engage domain experts to ensure that all legal and compliance requirements are met without compromising on the deliverables and objectives.

With the passage of time, laws and regulations have evolved and changed rapidly in terms of their jurisdiction, processes, engagement, applicability and interpretations. However, in the present scenarios, it is still an outreach, expensive and time-consuming as compared to the parlance of a fast-changing world. As such, no person or party would prefer to spend their time, resources, and money in courts to resolve their legal differences or conflicts which not only disrupt businesses but also hamper relations, goodwill and focused approach. Thus, it is important for all contracting parties to negotiate more balanced and agreeable contract terms, take suitable positions, allocate risks appropriately, diminish the likelihood of conflicts and litigations and reasonably distribute the liabilities and financial burdens. The list of possible issues, threats, risks, and problems related to legal, statutory and regulatory can be endless. In present and future businesses, the foresighted legal risk analysis is very important and crucial for the successful operation of the projects. Hence, engaging competent domain experts (like Legal, Risk or Contract), who can envision the probable areas of disputes and attached liabilities is strongly advisable. They can protect your interests, assets and goodwill and prove to be a boon for projects and organisations.

Though the ambit of applicable laws is very wide and construction contracts have numerous provisions, it does not mean that every situation, point, law, regulation, or provision can be equally risky. Some can be more tricky, risky, and complex as compared to the others. However, despite the best of intentions, plans and care practised by experts, there are chances that a conflict may arise during the performance of the project, as legal and project environment, situations and conditions are highly susceptible to changes.

Construction contracts are the main tool that defines the relationship between customers and contractors. The contract is the first level of defence for the parties involved to protect their interest and manage risks. Therefore, the contracts should be drafted and negotiated thoughtfully, and adequate care must be taken to allocate the risks and provide the mechanisms to mitigate risks. Deployment of professional legal, contract or risks specialists is the investment for the organisations, which can yield manifold dividends and returns in no time. Many companies lose due to a weak administration or management of their contracts during the project performance. Professional legal and contract professionals not only help to identify appropriate risks and take the right positions but also effectively add value in leveraging their legal benefits by substantiating the variations and claims through provisions in different legal frameworks.

Modern construction projects involve multiple stakeholders' agencies, which demand all to work together and are dependent upon each other operationally, technically, commercially and legally. To prevent any potential disputes or conflict, it is advised that the roles, responsibilities, and risks of associated parties and agencies are to be explicitly defined and mutually agreed, as all activities, actions and decisions can have financial implications on these agencies, as all the parties and agencies are committing their time and resources. That is why these relationships are to be documented in the form of a legally enforceable contract and do not depend on oral understandings or verbal assurances. The interest of all the contracting parties must be well-guarded, including of those persons who are taking the decisions or signing contracts on behalf of the party or agency. Projects and organizations may last longer but decision-makers, negotiators and staff may change anytime. This underlines the need for maintaining timely and correct records of basis, decisions, dependencies, and scope, etc.

First and foremost, the highest risk is to start the project work without having duly executed a written contract between the parties. It is of utmost importance that parties must commence the project activities only after signing a legally enforceable binding contract by the competent authorities of all the parties involved. Many contractors or suppliers may commence project works without the receipt of a purchase order or contract, anticipating that the customer shall issue the same later or may accept their proposal as it is, which may not be true.

The contract is a legal document and defines the rules of the game, so it is recommended not to mix (replenish) the emotions and relationships with the need for timely signing the contract documents. The contract works as a Bible between the parties and in case of any disputes, the parties (including courts) stick to the executed contracts. Projects may invoke a conflict at any stage. Thus, while starting the project work, all parties should be comfortable and confident of protecting their interest and working to achieve a **win-win** situation with the agreement.

The parties must expressly understand the underlined requirements and parameters of the project like the scope, schedule, budgets, quality, safety, and relevant legal and statutory aspects. The parties should demonstrate commitments to ensure that:

- Scope is clearly defined and shall not hamper in between,
- Schedule deadlines shall not slip,
- Approved budgets or contract price shall not overrun,
- Quality plans, standards and requirements are not compromised,
- Confidentiality, IPRs and Liabilities are not infringed,
- Human life or properties are not unreasonably jeopardised,
- Legal and statutory compliances are not violated.

Ultimately, the project objective is achieved without conflicts, ambiguities, heartburns and disputes. Therefore, the contract document should provide a framework to overcome the above aspects by defining the roles, responsibilities, obligations, and limitations of each party.

The legal relationship between the parties is not merely limited to contracts but also to the intent and conduct which is also equally important. Hence, proper contract management is as critical and crucial as the contract itself.

In the event a project requires the deployment of multiple agencies at a site concurrently, then there should be multiple contracts with different agencies and all the parties should be able to work in harmony and coordination without damaging the core objective of the project. It means, if there is a dependency of one party on another party, the same must be clearly laid down in the contracts so that any party is not unduly penalised for reasons not attributed to it. The contract should have a clear visibility to the promises and reciprocal promises.

When the main contractor is willing to participate in a tender or submit its proposal to their customer, the contractor may collate many offers from downstream vendors or sub-contractors for qualification or estimation of their final proposal. This stage is called selling, proposal, or bidding stage. Proposals received by the main contractor are used to prepare its final offer for submission to the customer. This is called "bid-to-bid offer" or "bid-to-participate offer." At this stage, the contractor itself does not have the order and thus is not in a position to give a firm commitment to down-line vendors or sub-contractors. These may be referenced as budgetary or estimation offers, whereas actual orders are awarded to the down-lines when the contractor wins the orders from the main customer. Furthermore, the main contractor may not award the order to vendors or sub-contractors based on their previous (bid-to-bid) offers and may seek further discounts or revised offers before concluding the order. So, this clarity must be sought by vendors or sub-contractors, as the validity of prices (including inputs costs) may lapse in a reasonable time. Nonetheless, the normal gestation periods for projects are relatively much longer than as required to finalise the products, transactions, or small orders. Thus, parties should ensure that price validity is thoroughly agreed because it shall then require to bind the downward suppliers or agencies.

In spite of well-defining the contracts by experts, there are probabilities that issues, conflicts, risks or discrepancies may arise in the future which can cause legal problems and force the parties to seek contractual positions or legal remedy.

Some of the common issues faced by the construction teams are related to: –

a. **Type of contracts** – turnkey, fixed-price or bill-of-quantity based, time and material, etc.

b. **Subjective terms** – "fit for purpose" or "to the full satisfaction of the engineer," etc. avoid such generic or subjective obligations – restrict to specs and contracts.

c. **Vagueness** – Gaps on correctness and completeness of prices (with inclusion and exclusion)

d. **Changes & variations** – related to the design, scope, assumptions, responsibilities and reciprocal promises

e. **Defects or rejection** (work or material) related to makes, sources, types, workmanship, standards, tests, inspection, acceptances or certification

f. **Delays & defaults** – by either party to discharge their obligations in the time and manner agreed or required

g. **Risk purchase provisions** – ideally liability should be within the cost incurred over and above the contract price by the other party

h. **Inter-dependencies** of other parties or agencies working in parallel at the site

i. Suspension, work stops, cancellation of work performance

j. ROW, social or political issues affecting performances

k. **Unclear or ambiguous contract** – plans or provisions or changes thereof

l. **Back-to-back terms** – rolling down from end-user to contractor to sub-contractor

m. **Force majeure events** – occurrences, reporting, claims and settlements

n. Warranty/defect liability provisions

o. Change in law, etc.

Dale Carnegie correctly said that *"**Each party should gain from the negotiation**."* In all said and done, contract negotiation is an art. There are instances in the industry, where the parties first agree on a broader scope or even the contract price. Later, they start the detailed contract discussions, which open up scope for significant changes or differences between parties.

At times, the customer demands very onerous or unfair conditions and then the parties fail to negotiate and arrive at mutually acceptable contract terms. Such situations complicate the execution as well as the relationships, as most of the contract terms can have direct commercial impacts on the project feasibility triggering from various approvals, permissions and compliances. For example, some customers may require registration of contracts with competent authorities, which involves a cost of approx. 0.5% (half a percent) of the total contract prices or assets involved, which may not have been envisaged by the contracting parties in the beginning. If the same are not complied with, it can be a violation.

To minimise the risks arising out of contractual or legal issues, the following provisions can be clearly defined or included in the contracts: –

a. **Scope of works** (SOW) – Scope, design, specifications, exclusions, changes, division of responsibilities, assumptions, etc. related issues are the common cause of conflicts in construction projects. One of the key reasons is the different understandings, interpretation and references related to scope, technical specification, and contract provisions. There are instances where a contractor uses different types of designs, methods, specifications than what was envisaged in tender documents. Such a confusion occurred due to conflicts in the tender and contract documents or gap in understanding.

Some customers may not fairly disclose information at the time of the tender or award of contract. For example, customers intimate that the targeted construction site is clean, flat, unfettered, normal soil, and rectangular in size. Accordingly, the contractor assumes the same in its layout, design, BoQ, planning, scheduling, budgeting, and risk provisions. However, during actual field surveys or assessments, the contractors found major deviations, which significantly impact the tender planning and assumptions. Since the contractors agreed on fixed-price contracts, this may become difficult to revise the same easily or cause unnecessary conflicts.

Likewise, a contractor assumed the availability of the encumbrances-free site, unfettered access and approach to the site, no electrical lines, structures, trees, *nallahs*, etc. However, if these impediments exist at the site, then it can shatter the execution plans from the beginning itself, as the removal of such hurdles cost time and money. The contract should clearly define the scope of work for the parties and distribution of responsibilities, with boundaries, battery limits and inter-dependencies, so that the parties can perform their respective scope of work in the time and manner agreed.

Additionally, the contract may spell out as to how the contractor will handle unforeseen events, ensure completeness of work with the required level of quality standards. In case of incomplete scope definitions, problems may arise during construction and invoke dispute concerning the scope or quality of work. Such disputes shall delay the construction work and impact the project's financials. It is possible to prevent such exposure by properly defining the scope, prior field assessments, and seeking concurrence of parties. All relevant documents are then made an integral part of the contract and the order of precedence is also documented. Applicable codes, standards, specifications and laws are clearly defined in the contract, so as to eliminate ambiguities or conflicts.

b. **Work acceptances criteria** – The customer's representative periodically certifies the measurements of the works completed by the contractor as per contracts, billing break-up or milestones agreed. Based on work certification from sites, the contractor raises invoices (payment claims) by annexing the other documents. Most contracts require that all payment claims should be substantiated with relevant documents like material inspection certificates, dispatch authorisations, lorry receipts, goods receipt notes, tax invoices, etc. Upon receipt of payment claims, the customer pays the undisputed amount to the contractor, as per agreed terms of payment. Disputes may arise when there is no clear milestones, acceptance criteria, completion criteria, certification processes is defined or no authority granted to the person at site for certification.

Some contractors may seek certification of work or activities which are incomplete, excess of actual work, or defective works, thereby opening up the scope of mistrust and disputes. If the customer finds that the contractor has estimated or claimed undue works that do not have adequate merits for payment, then customer not only holds the payments but also reviews as a breach of contract. If such issues remain unresolved, then it leads to unnecessary conflicts, disputes, and litigations. Thus, parties must agree on processes, acceptance criteria and documents to demonstrate completeness or acceptances of each line item. As an extra precaution, many parties may engage an independent third-party inspection agency, who inspects the measurements and recommends payments.

Due to weak project planning, some contractors may supply the high-value material at the early stage of the project, to generate revenues, rather than what is actually needed at sites. They also push for the items what are billable first and accord low priorities to the material or works which are non-billable or free part of the contract. This creates a lot of conflicts, confusions, and issues between the parties later because contractors pretend to be left with no money and lose interest. So, proper project management practices are to be ensured to make a win-win project.

c. **Price basis** (fixed or variable): – It must be explicitly defined as to how the contract prices are determined. If this is firm and fixed-price contract, then complete risks and rewards of uncertainty rest with the contractor. Fixed-price-contracts give a high degree of cost certainty to customer. Such contracts provide limited opportunities to the contractor under which it is entitled to the changes. Fixed-price-contracts include all planned costs, contingency costs and mark-ups for overheads and profits. Generally, firm and fixed prices-based contracts are relatively expensive for the customer, but then the customer is absolved from the uncertainty and vulnerability of the project and market conditions. If it is not documented correctly in the contract, then it unnecessarily becomes a conflict when a change arises. Such situations can cause disputes, especially when costs hit negatively.

Whereas, if the contract is based on variable prices, then risks and rewards of uncertainty rest with the customer. In case of variable price-based contracts, the parties must define the base date or reference date of the prices, indexes of references (commodity and labour rates or foreign currency), formula for calculation of variations, limits and ceilings at which variation shall be considered, treatment of taxes, duties and other levies, etc. to be agreed upon in the contract and not leaving the scope of conflict for the later stage. Such a contract should also specify the mark-up margin or fees to be considered by the contractor while lodging its claims.

d. **Payments** – Contacts must spell out clearly the terms of payment as to when the contractor shall be entitled to receive the payments for its obligations performed. For example, if it is supplying any equipment, as to when he will receive the payment against the supplied equipment or agreed Incoterms. Whether payment will be due against dispatches, delivery at port, delivery at project site, after erection or commissioning of the equipment. Normally the equipment payments are made in multiple tranches. The major portion of the supply payments are made as per agreed Incoterms, and the rest is withheld until project completion. Payment terms play a vital role in managing the project cash-flow. As per agreed payment terms, the contractors can then negotiate the terms with their downstream sub-vendors or sub-contractors. Some contractors agree on back-to-back basis payments with their downward partners like 'pay when paid' or 'pay if paid' terms. This is not only a support in managing their cash-flow but ensuring that the supplies (or work) performed by downward sub-vendors or sub-contractors get accepted by the ultimate customer. Major issues crop up while retention payments are released; so clear terms, documents and requirement must be laid down, to smoothly realise the retention amounts, when due.

e. **Schedules & timelines** – *Time is money.* Timelines (phases, milestones, stages, activities) are crucial for the success of the project. At times, the customers expressly define stringent provisions like '***time is the essence of the contract***.' It means the customer does not expect any unreasonable delays and recover the loss or damages due to project delays. Schedule-related risks are common. The contract should clearly spell out the start dates, end dates and duration of each phase, milestones, and key activities. Commonly, construction contracts define that start dates shall commence upon the fulfilment of certain condition precedents. After the achievement of condition precedents, the customer may issue the written document called Notice to Proceed (NTP) or deemed start-date or coming-into-force conditions. Such pre-conditions may include (a) Signing of the contract by both/all parties; (b) receipt of advance payments by the contractor; (c) preparation of the detailed project schedule; (d) providing site access to the contractor or other similar conditions. Certain contracts define the clear hard dates, regardless of the fact whether pre-conditions are met or not, but end dates are sacrosanct, and delays are subject to punishments, which may include levying liquidated damages, penalties or invoking termination, etc. Start-date and end-date of the project must be clearly reflected in the schedule with inter-dependencies so that the project delays (if any) can be fairly ascertained and attributed to the defaulted party.

f. **Liquidated damages (LD) & penalties** – *Time is an essence.* Delays can affect project schedules and procrastinate to achieve the intended objectives. Contractors may seek time extension and/or damages for the events over which they have no control or responsibility. However, if the delays are not attributed to the customer, the contractor has to compensate to the customer and may have to bear liquidated damages as contractually agreed. All delays are not qualified under *force* *majeure* and the contractor may not be excused for the delays. Even some *force majeure* conditions provide exclusion; it means, delays are not excused or exempted and accountability rests with the contractor. Besides levying LD, consequences of delays can be very expensive due to inflations and fluctuations and other costs of prolongation can hit the project budget severely. Ultimately, the contractor bears the compensation for delays and the burden of prolongation cost. One of the solutions is to agree on compensable events (or activities) or non-compensable events (or activities).

Most projects may face situations of delays; therefore, such events should be visualised in advance and amicably resolved before converting into actual losses, conflicts, or litigations. Many organisations may deal with this issue by making the contract more lucrative by making provisions for bonuses or incentives for early completion, thereby not only eliminating the risk of delays and liquidity damages but also exploring the opportunities of early completion and using the facilities.

Project performances are usually delayed beyond the stipulated time-frame. So, the contract should define such situations and recourse to resolve them. The parties should agree on the rate of liquidated damages to be levied with maximum ceiling. Normal rates of LD vary from 0.50% to 1.00% per week with a cap upto 10–20% of the total contract price. Many parties may agree to the rates or limit for the unexecuted portion of the contract and not on the total contract value. Some contracts may define LD on multiple milestones like submission of drawings, x% material supplies, x% civil work, erection works and/or commissioning, handing-over or any other key milestones. In addition to the delay LD, the customer may demand penalties for losses or violations related to production, compliances, performance guarantees or other reasons. Such conditions and calculation formulae must be agreed in the contract, to minimise the risk of LD and penalty.

However, the contractor (and its downstream parties) must thoroughly evaluate the feasibility of executing the project on time and monitor continuously to meet the schedule. Nonetheless, the contractor should also make adequate documentation to substantiate its case by ensuring that the reasons for such delays are not exclusive and not solely attributed to it. In case the delays are attributed to the customer or situations beyond reasonable control, the treatments and evidence should be maintained and presented to prevent such high risks of LD. Organisations should consider certain risk provisions or contingency on account of timelines and based on their project experiences or capabilities. Contracts may have inherent conflicts within and it is difficult to find an express remedy for such defects or defaults. For example, parties agree for the project timelines and liquidated damages provision and define it as "genuine pre-estimate damages that customer shall suffer resulting from the delay in the project" and at the same time it defines that "time is an essence" and furthermore it specifies "if the customer does not actually suffer losses, no liquidated damages shall be payable by the contractor." This is based on 'no-harm no-foul' principle. This causes big confusion and conflict and becomes a source of litigation. Similarly, many times, contractors agree with fixed-price-contracts with low or no opportunity of revisions. But they agree to the contract provisions with their down-line sub-contractors due to their limited scope involved on different terms, which causes concerns and conflict. Thus, special caution is to be taken while concluding the contracts. Similarly, many contracts state that LD is the 'sole and exclusive' remedy for delays, but at the same time contracts specify the open-ended clauses of termination for convenience and risk-purchases, which may be interpreted contrary to the 'sole and exclusive' remedy.

g. **Force majeure** – Contracts should define what is considered as *force majeure*. What is covered and what is excluded from its definition. Not every 'Act of God' can be considered as *force majeure*. For example, events like high temperature, heavy rains, land sliding in hills, etc. may be excluded from *force majeure*, as these are regular phenomena at certain areas or geographies. Further, it should be laid down in the contract as to what actions need to be taken upon occurrence of *force majeure* event and what reliefs can be considered based on the merits of the situation. Many contracts require that a written notice be provided to the customer within a specified time. Most contracts provide a relief in terms of a commensurate time extension and not cost benefits in lieu of *force majeure* impact. Further, any damages to the equipment and material due to *force majeure* event also generally excluded from the performance warranties of suppliers. Such occurrences may invalidate the manufacturer's warranties to the damaged material and equipment. So the risk of replacement or repair may directly be the contractor's responsibility if the same is under their storage, care, custody and protection.

h. **Right to suspension** – Many contracts force the contractors to perform their scope of work without stopping or suspending the activities regardless of the defaults or situation. It means, a contractor cannot stop or suspend the work, once started. No contractual right is provided. This is quite common in government contracts, where contractors are not given any contractual right or relaxation from the customers. In case there is a compulsion to suspend or stop the work, it shall be exclusively on the risk and cost of the contractor. Therefore, it must be clearly understood or well-negotiated during the formation of the contract that in case customer defaults or fails to discharge its timely obligations in the manner prescribed in the contract, the contractor has the right to stop the work. However, in case suspension rights are not provided and there is a compelling reason, the contractor should consider how it is going to deal with the time-lapse and idle costs, as the clock will keep ticking.

i. **Right to termination** – Many contracts force the contractors to perform their scope of work without terminating or cancelling the work regardless of the defaults or situation. It means, the contractor cannot terminate the work, once started. No right is provided, whereas the customer may have the rights to terminate the work with or without assigning a reason. Many contracts provide the rights to the customer to terminate the contract for convenience and get the work done through a third party on the risk, cost and consequence of the existing contractor. Sometime, projects may become unviable for the contractor to continue but it cannot terminate. In the absence of suitable provisions in the contract, the contractors are compelled to take legal recourse in the courts of law. The legal procedures are expensive and time-consuming also. However, the cost and consequence of the termination of the project can impact heavily on the party with financial losses and loss of reputation. Therefore, the contractor should negotiate for a balanced contract with equal rights of termination, at least in the case of major customer defaults or frustrations in discharging their obligations in the time and manner provided in the contract.

j. **Acceptance criteria** – Contracts to cover the acceptance criteria for various tasks and activities to be performed by the contractor.

   i. <u>Drawings and documents</u> – parties should converge on the list of drawings, documents, specifications to be submitted for approval alongwitth communication plan, response time (service level agreements) and frequency for reverting on documents (approval/rejection/return with comments).

ii. Material Quality – parties should agree for approved makes, approved supplier or vendor list, source or origins, manufacturer's quality plans, tests (routine, functional or type tests), and inspection plans, inspection agencies. Also agree on categories, codes, standards, types and places of inspections (factories, distribution centres, labs, or project sites), dispatch authorisation processes, SLAs etc. In case third-party agencies are deployed for inspection, then finalise the agency, skills and experience of the inspectors and their costs should be agreed. Also define the exact type tests or routines tests to be carried out for the material and as to who shall bear the cost and time.

iii. Service Quality – For services related to civil, mechanical, electrical, logistics, etc., the parties should agree on the field quality plan, raw material (civil works), test labs, inspection methods, audit process, SOPs, trainings, checklists, reports, formats, etc.

iv. Site Works – List all the site works, methodologies, packages, stages, frequency of supervision, inspections, work measurement and certifications, SLA for revert to be agreed so that there is no undue delay or conflict during the project execution.

v. Change & Variations – Changes are inevitable in projects. Changes and variations processes, communication and approval matrix, authorisations, service level agreements and order amendment procedures should be agreed.

vi. Stores & inventories – Clear processes to be laid down for all material receipts, acceptances, rejections, storages, stacking, issuances, returns, spares, wastes, and scraps disposal related. Also agree on the frequency of material reconciliations and treatments of damages, shortages, etc.

vii. <u>Handing-over/Taking-Over</u> – As all projects are temporary and have to reach an end at some point of time, the parties must agree on the complete handing-over-taking-over (HOTO) process, requirement of documents, spares, tests, punch-lists or any other requirement so as to perform the HOTO activities smoothly. As many times, customers may not have adequate staff or operations team to take over the control or possession of the executed projects.

k. **Warranty & defect liability** – Generally all contracts define warranty provisions to ensure that the products, services and projects conform to the agreed specifications and stand fit for the purpose, and in case of an apparent defect, deficiency, or damage during operations, how it is going to be rectified, remedied or dealt with. Therefore, the contracts should explicitly stipulate the start and end stages (or dates) of warranties and duration for which the warranty obligations shall be honoured by the contractor. Often organisations agree for a warranty period ranging from 12 to 60 months, depending upon the customer requirements and accordingly consider the cost of warranties. In case a contractor agrees to the extended warranty with its customer, they must ensure that their warranty commitments are fully backed up with back-to-back commitments from their sub-suppliers or sub-vendors. It should expressly define exclusions (if any) in the warranty clause. Many contract provisions define the limitations or constraints of the warranty, under which contractors are not liable to perform warranty obligations or even may make the warranty null and void.

Due to occurrences of defects or deficiencies in the products, material, systems, the customer may be deprived off the committed production, generation, performance or usage of facilities; then the contract should define how such downtimes or losses of use shall be treated between the parties. The parties may agree upon the response or rectification time of the defects and restore the product in service, so as to minimise the customer loss. Contractors may agree to extend the warranty periods (beyond the original warranty period) from the time when defects are rectified. It is also strongly advisable that parties must agree on the minimum spares to be maintained at facilities/plant for immediate rectification or replacements. Further, many contracts require the coverage of a latent defect period, which exceeds the warranty period. Most of the time, it becomes very painful for the suppliers or contractors, when they agree to the evergreen warranty provisions, where the warranty keep extending or restart with every rectification of defects.

l. **Battle of forms** – Construction contracts involve a large number of documents to be annexed as an integral part of it. This may include tender documents, circulars, amendments, proposals, minutes of meetings, drawings, specifications, general conditions of contracts, special conditions of contracts, purchase orders, release orders and other exhibits and annexures. There are regular phenomena that cause conflicts between multiple documents and versions. So, conflict arises as to which documents prevail or supersede in case of a conflict. Thus, contracts should define the list, hierarchy and order of precedence of various documents, so as to de-risk any ambiguity and take a faster decision in the interest of the project.

m. **Liabilities & indemnities** – Limitation of liabilities clause, stipulated in the contract, is actually a safety net for parties. It protects the contractor against any claims or lawsuits filed by the customer or vice-versa. It caps the amount of potential exorbitant damages or losses to which a company is exposed when the contract is breached or defaulted. Generally, parties agree on 100% as the liability cap (equivalent to the contract value). However, many parties may define higher X caps in the contract, although liabilities may put a cap on the direct liabilities. However, parties agree on certain carve-outs, which are not covered in the stipulated ceiling amount. Such carve-outs may include infringement of gross negligence, wilful misconduct, IPR, confidentiality, indemnities, penalties, and/or other applicable laws, etc. Generally, liability, indemnification and insurance clauses are defined in the contract for the purpose of risk allocation. Insurance transfers the risks from the customer or contractor to the Insurance company. Indemnification transfers the risk between the parties to the contract. Limitation of liabilities basically controls or prevents the transfer of risk between the contracting parties. Many great companies insist on the incorporation of this clause and can be non-negotiable or deal breaker.

n. **Indirect, consequential, punitive & special losses & damages** – Such losses or damages in the construction industry have been frequently contemplated among project teams, contract experts, lawyers, and courts which causes ambiguity, conflicts and disputes. Generally, loss of profit, loss of generation or production, loss of revenue, loss of opportunities, loss of goodwill, etc. are considered to be indirect or consequential losses for which parties agree to exclude from the liabilities of contracts. However, there are frequent ambiguities whether the loss is indirect or direct. One of the good ways to de-risk such conflicts is to clearly identify in the contract as to what is excluded before executing the contract. Many great companies insist on the incorporation of this clause and can be non-negotiable or deal breaker.

o. **Intellectual property rights (IPR)** – IPR is the legal right resulting from the human intellect which includes concepts, innovations, creativity, trademarks, copyrights, patents, industrial designs, trade secrets, brands, etc. Some of the inventions could be intangible, very expensive and could be the foundation of businesses. In the current globalised information and knowledge economy, every IPR runs the high risk of infringements. However, there is a lack of awareness for IPR and insufficient regulations to protect the risk of IPRs. IPR registrations can extend an assurance and guarantee to the original owner of its rights and benefits. Organisations develop, evolve, and improve their intellectual property management continuously and take suitable means of prevention of infringement risks, protect their innovations, and promote continuous innovation and development. So, appropriate provisions should be incorporated in contracts specifying the purpose, period and limitations of providing any IPR rights to customers.

p. **Governing laws & jurisdiction of courts** – Choice of law specifies the laws that will govern the contracts and achieve certainty. It also supports in assigning proper meaning, interpretations, enforcement, and validity regarding the rights and obligations of the contracting parties. Generally, parties are free to opt any governing law. Jurisdiction of courts defines the relevant courts or arbitration tribunal that will hear any disputes arising out of the contracts. Parties must decide on the governing laws and applicable acts in case of disputes arising from the contract. Parties also define the arbitration process and place of courts, etc. in the contract.

q. **Dispute resolution** – Conflicts, differences and disputes are common in projects. So, defining dispute resolution or dispute settlement mechanism guides the process to resolve disagreements between the parties through negotiations, conciliation, mediation, expert determination before such disputes are referred or escalated for arbitration or litigations. Construction contracts should define dispute resolution clause in a clear and concise manner and anticipate future problems that may arise. It should clearly define what is considered as a dispute, scope of this clause, as well as when and how it can be triggered. It should also define if this dispute resolution fails, how the parties can escalate or proceed further. In case of arbitration requirement, parties can determine the number of arbitrators, but it should not be an even number. The parties may also choose to continue the performance of the project work during the dispute.

r. **Changes, variations & claims** – Projects are executed in very uncertain, dynamic and volatile situations and there is a high probability of changes that may crop up anytime. It is a disbelief or myth that a project shall not have any changes whatsoever during performance. Even well-defined and negotiated contracts pass risks and responsibilities to contractors, which has the potential for changes, variations and claims. These changes can affect the schedule, budget, scope, specifications or other plans or parameters of the project. Sometimes, such clauses are not incorporated in the contract, with a consideration that the contractor shall inflate the contract prices or unnecessarily create changes to make money or delay the project. But upon occurrence of crucial changes, the contractor holds the construction and affects work at the site that spoils the relationships and impacts the project. If it gets worse or exceeds the contract price ceiling, this may lead to litigations as well. The easiest way to circumvent this risk is to include the Change and Variations provision in the contracts but negotiate the changes more thoughtfully. The parties should properly discuss and define the methods and processes for initiating, reviewing, processing, substantiating changes and reflecting the impact thereof. Ultimately, one cannot formally close the contracts with pending changes, variations or claims.

s. **Completion & closures** – Most construction defects are brought out during the completion stage, which leads to conflicts and disputes. The parties even forego good works, abuse the relationships, cross the professional boundaries, and start blames and allegations of negligence or breaches. Completion, closure and handing-over-phase is the most complex phase in any project. Some projects even take over 20% of the total time in closing projects and completing punch-points, rectifying defects, or completing the documentation. The parties should clearly define the handing-over process, deliverables, procedures, checklists, documents, and other spares/consumables, etc. so as to make a smooth completion and exit from the project. Many customers may list specific activities or define milestones to demonstrate the completion. Even contractor retention payments are also linked with completion, which is released after achieving completion deliverables. This may also require submission of performance bank guarantees and submitting various undertakings, etc.

Concluding, as a contractor, it is imperative that it is difficult to manage so many diverse situations and circumstances; therefore, all the relevant provisions related to the above issues or more issues based on individual, industry, domains, experiences and expectations must be taken care of while drafting and negotiating the contracts. Thus, a written contract document is absolutely essential enumerating the obligations, requirements, expectations, and relationships between the parties. This will facilitate to achieve the objectives, meet financial goals and protect their interest for long and mutually beneficial relationships so that the parties can enjoy planned profits, avoid unwarranted costs, and avoid conflicts. It is impossible to visualise and mitigate all futuristic unforeseeable occurrences that may arise; however, with a limited knowledge, wisdom, learning and experience, the organisations can certainly make impactful contracting relationships for a long-lasting win-win situation.

Abbreviations

ABG – Advance Bank Guarantee

ASL – Approved Supplier List

AVL – Approved Vendor List

BBU – Billing Break-Up

BG – Bank Guarantee

BOCW – Building & Other Construction Workers

BOQ – Bill of Quantities

CEA – Central Electricity Authority

CEIG – Chief Electrical Inspector to Government

CFT – Cross-Functional Team

COC – Certificate of Conformance or Certificate of Conformity

CPBG – Corporate Performance Bank Guarantee

COPQ – Cost of Poor Quality

DOR – Division of Responsibilities

EAR – Erection All Risks

EOT – Extension of Time

EMD – Earnest Money Deposit

EPC – Engineering, Procurement & Construction

ESG – Environment, Social & Governance

FAT – Factory Acceptance Test

FFP – Fixed Firm Price

FQP – Field Quality Plan

GDP – Gross Domestic Production

GFC – Good For Construction

GoI – Government of India

GRN – Goods Receipt Note

ERP – Enterprise Resource Planning

HSE – Health, Safety & Environment

HOTO – Handing-Over-Taking-Over

ILO – International Labour Organisation

IPR – Intellectual Property Rights

JIT – Just in time

JV – Joint Venture

LC – Letter of Credit

LD – Liquidated Damages

LOTO – Lock Out/Tag Out

LSTK – Lump-sum Turnkey

NTP – Notice to Proceed

MDCC – Material Dispatch Clearance Certificate

MEP – Mechanical, Electrical & Plumbing

MQP – Manufacturing Quality Plan

MRN – Material Receipt Note

NTP – Notice to proceed

OEM – Original Equipment Manufacturer

PBG – Performance Bank Guarantee

PLC – Project Life cycle

PM – Project Management/Project Manager

PMI – Project Management Institute

PMO – Project Management Office

PPA – Power Purchase Agreement

PPE – Personal Protective Equipment

PTW – Permission to Work

QAP – Quality Assurance Plan

QMS – Quality Management System

QR – Qualification Requirement

RACI – Responsibility, Accountability, Consulting & Information

RCA – Root-Cause-Analysis

RFP – Request For Proposal

ROW – Right-of-Way

RMP – Risk Management Process

SCM – Supply Chain Management

SLA – Service Level Agreements

SOP – Standard Operating Procedure

SOW – Statement of Work or Scope of Work

SPOC – Single Point of Contact

SPV – Special Purpose Vehicle

TCO – Total cost of operations

VUCCA – Volatile, Uncertain, Complex, Chaotic & Ambiguous

WBG – Warranty Bank Guarantee

www.ingramcontent.com/pod-product-compliance
Lightning Source LLC
Chambersburg PA
CBHW030931180526
45163CB00002B/525